High Plains Horticulture

High Plains Horticulture
A CULTURAL HISTORY

JOHN F. FREEMAN

UNIVERSITY PRESS OF COLORADO

© 2008 by the University Press of Colorado

Published by University Press of Colorado
1580 North Logan Street, Suite 660
PMB 39883
Denver, Colorado 80203-1942

All rights reserved
First paperback edition 2023
Printed in the United States of America

 The University Press of Colorado is a proud member of the Association of University Presses.

The University Press of Colorado is a cooperative publishing enterprise supported, in part, by Adams State University, Colorado State University, Fort Lewis College, Metropolitan State University of Denver, University of Alaska Fairbanks, University of Colorado, University of Denver, University of Northern Colorado, University of Wyoming, Utah State University, and Western Colorado University.

ISBN: 978-0-87081-927-8 (hardcover)
ISBN: 978-1-64642-569-3 (paperback)
ISBN: 978-0-87081-983-4 (ebook)

Library of Congress Cataloging-in-Publication Data

Freeman, John F. (John Francis), 1940–
  High Plains horticulture : a history / John F. Freeman.
     p. cm.
  Includes bibliographical references and index.
  ISBN 978-0-87081-927-8 (alk. paper) — ISBN 978-1-64642-569-3(alk. paper) — ISBN 978-0-87081-983-4 (ebook)
  1. Horticulture—High Plains (U.S.)—History. I. Title.
  SB319.2.H54F74 2008
  630.978—dc22

2008024777

Contents

Preface vii
Acknowledgments xi
Introduction 1

1. Horticultural Beginnings 7
2. Trees for the High Plains 19
3. Horticulture for Home and Community 33
4. Toward "A New Phase of Civilization" 51
5. Science and Its Application to Horticulture 63
6. Creating Home on the Range 85
7. Limits of Dry-Land Horticulture 107
8. Forging New Paths in Ornamental Horticulture 129
9. Collecting and Creating Hardy Plants 151
10. Federal Engagement in Horticulture 171
11. The Cheyenne Horticultural Field Station 195
12. Horticulture and Community 227

Postscript 243
Bibliography 249
Index 261

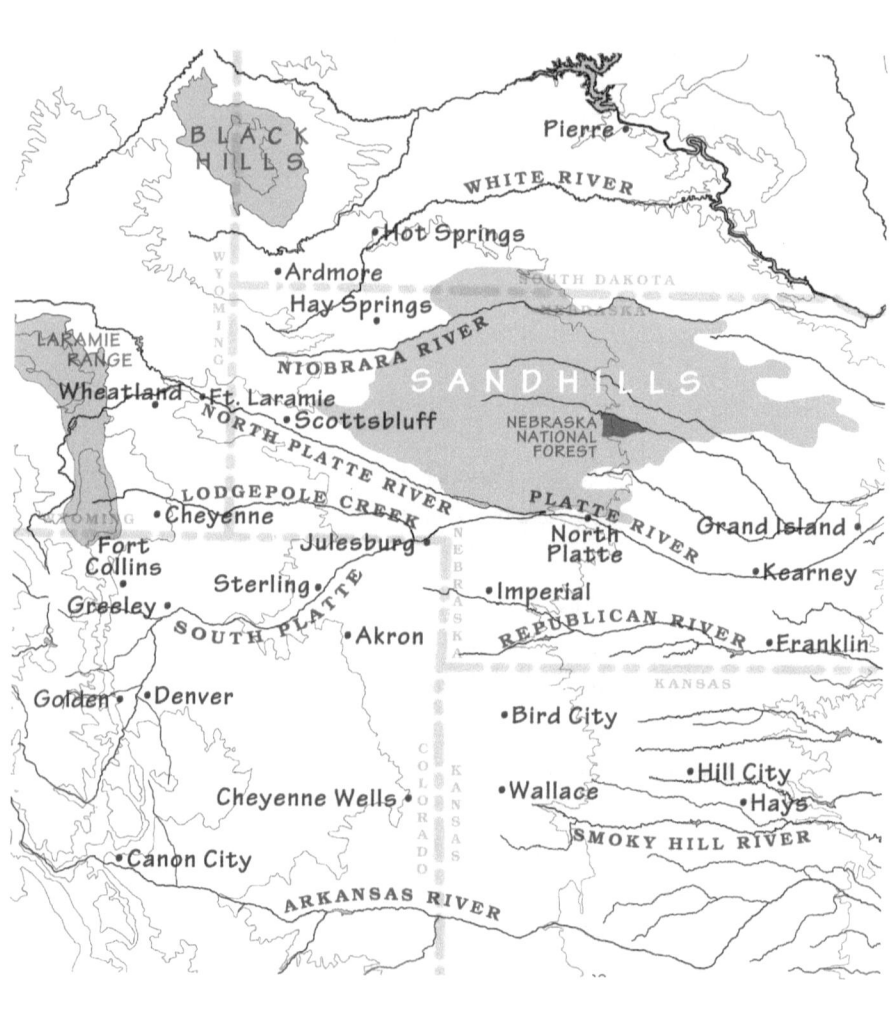

Preface

The civilizing role of horticulture is part of the settlement story of the High Plains that has yet to be a subject of special consideration. The significance of this topic may be most readily explained by telling how it originated and developed in my own mind.

In 1954, as an eighth grader, I was driven across the plains of Kansas, Colorado, and Wyoming. After spending the summer in a primitive cabin in a remote location in southeastern Wyoming, I took my first train ride alone, from Laramie across the plains of Nebraska on to Chicago and the East Coast. Even at that young age, I had found the High Plains awesome and exhilarating, not lonesome or bleak, and I hoped someday to put down roots there.

While attending university in the Midwest, I had the good fortune to be able to spend holidays and summers in Colorado. I remember vividly the first welcoming whiffs of sagebrush, as well as the oases of small towns and farmsteads, as I drove west through Kansas or Nebraska. Indeed, my academic interest in French rural history, both civic and ecclesiastical, seemed perfectly compatible with what I saw on the High Plains. After moving from southern California to Wyoming in 1971, at Rock Springs, then a booming energy town and not an apparent garden spot, my first question was: Where are the trees? The answer, invariably, came back: none of us thought we were going to stay here very long, so we did not plant trees.

Some years later, as founding president of the Wyoming Community Foundation, I became convinced that building endowments that would last forever and planting trees that would survive the harsh climate were two sides of the same coin: creating permanent communities. Among the Community Foundation's first grants were monies for the purchase of trees for several Wyoming towns. Furthermore, at the urging of Wyoming's then first lady, Jane Sullivan, the Community Foundation supported a number of projects and activities to improve community appearances. The premise, of course, was that attractive communities not only help increase community pride but also help attract desirable new businesses.

In the summer of 2001, almost by chance, I learned about the United States Department of Agriculture's Cheyenne Horticultural Field Station that had occupied a once-treeless 2,200-acre plot just west of that city. Established by act of Congress in 1928, its mission was to aid horticultural development in all aspects throughout the High Plains. In 1972 that mission shifted to grasslands research, reflecting a stronger commitment to ranching and related forage crops. Now efforts are ongoing to restore the station's arboretum, reflecting the interests of an increasingly urbanized population. The Horticultural Field Station thus remains institutionally convenient to our story.

Contrary to lingering public opinion—especially strong where I live—that nothing grows where drought is ever-present, temperatures are extreme, and the wind rarely stops, I have had the privilege of cultivating my own vegetable and fruit garden, with surprisingly good results over a number of years. That is not to say that this is a how-to guide, although today's High Plains gardener will learn, at least generally, what grew in the past and what did not. Nor is this a study of commodity farming and open-range ranching, although both are important because they complement horticulture and sometimes compete with it for water. Nor again is it about horticulture as secular religion, although horticulture as restorative remains admittedly attractive as it has throughout history, most notably since the patricians of ancient Rome first cultivated their own gardens. The reader is forewarned that this study purports to be a cultural, not a scientific or technical, treatment of horticulture, although I hope it is well-grounded on both the science and the applications of science practiced during the respective historical periods.

The reader should also know at the start that the notion regarding the civilizing influence of horticulture on the High Plains derives from my study and admiration of the learned French agriculturists of the eighteenth century. Actually, one need go no further than defer to one of their correspondents, our own Thomas Jefferson, overseer of the Louisiana Purchase, of which the High Plains were part. It is well-known that Jefferson was an enlightened farmer, that he viewed farming and, by extension, horticulture to be the most ennobling profession and the community of farmers to be the "nursery of steady citizens." While not mentioning the High Plains by name, Jefferson had written about the gradual shades of improved living from the untamed condition along the Rocky Mountains to the tamed condition in Atlantic seacoast towns, illustrating in one snapshot "the progress of man from the infancy of creation to the present day." In 1824, when Jefferson wrote these lines, the High Plains had yet to experience his "march of civilization."

Jefferson's hope for the gradual improvement of the human condition, not to speak of his intellectual curiosity, clearly had tempered his instructions to Lewis and Clark on dealing humanely with the aborigines they would encounter. The resistance of the Plains Indians to the "march of civilization" raises an overarching moral question that goes beyond whether their displacement was justified or inevitable: Can sophisticated, technically advanced use of the land be defended as equally virtuous to more primitive or pre-pastoral use of the land?

Clearly, as a resident of the High Plains, I am inclined to believe American settlement of the High Plains was a good thing. And I am grateful to the settlers who moved here, cultivated, and embellished their surroundings with the amenities of civilization. The history of their horticultural endeavors lends credence to the thesis that, slowly but surely, we have been learning to accommodate ourselves to the limits of our land.

Finally, as an immediate and practical matter, there is an advocatory rationale for telling this particular story. We have reached the point, in the early twenty-first century, when the overwhelming majority of us, even on the High Plains, have lost day-to-day touch with the soil. And while we likely will never get back to the family farm, much less to the struggle to survive the elements, it might just be feasible and certainly desirable for us to cultivate our own gardens, no matter how small, even if such activity borders on the sentimental. Similarly, while we may never get back to the idyllic community, we can make our surroundings more attractive, thus more livable.

Acknowledgments

As the notes and bibliography illustrate, my research was based primarily on regional sources. Thus, I am most grateful for the assistance of archival and library staff members at the Colorado State University, University of Nebraska–Lincoln, South Dakota State University, and University of Wyoming. In particular, I wish to recognize Dee M. Salo, interlibrary loan librarian at Wyoming. Also, I thank Joe Becker, Western Kansas Agricultural Research Station, Hays; Peggy Ford, Greeley Museums; Ronald K. Hansen, Horse Creek Studio, Laramie; Sue Lowry, Fort Laramie National Historic Site; and the staff of the High Plains Grasslands Research Station, Cheyenne.

For suggestions and encouragement, I am pleased to acknowledge in particular Mark Hughes, community forester, Wyoming Forestry Division; Scott Skogerboe, plant propagator, Fort Collins Nursery; and Shane Smith, director, Cheyenne Botanic Gardens.

For critically reading the manuscript, I am indebted to Lara Azar, former press secretary to Wyoming governor Dave Freudenthal; James R. Feucht, professor and extension horticulturist emeritus, Colorado State University; Glyda May, retired rancher; and Roger L. Williams, professor of history emeritus and affiliate of the Rocky Mountain Herbarium. Finally, I acknowledge with both appreciation and affection my many friends and acquaintances working in the voluntary, private, and governmental sectors to make life better for all on the High Plains. I hope this book will not disappoint them.

High Plains Horticulture

Introduction

The story of horticulture on the High Plains began very inauspiciously. In 1806, Zebulon Pike, the first known American explorer to cross this region, reported on "barren soil, parched and dryed up for eight months in the year that, in time, would become as famous as the sandy desarts of Africa." In 1820, Major Stephen Long and his fellow explorer, botanist Edwin James, reported that the region was "almost wholly unfit for cultivation" and "an unfit residence for any but a nomad population." In 1846, a Kentucky journalist recorded that western Nebraska was "uninhabitable by civilized man"; and in 1849, historian-horticulturist Francis Parkman described the entire country from the Missouri River to the Rocky Mountains as a "barren, trackless waste."[1] Thus began the tradition of describing the region as the Great American Desert, both on early geographic maps and in later folklore.

Taking pecuniary advantage of the admittedly harsh climate of the High Plains, Bill Nye, editor of the *Laramie Boomerang* in the 1880s, entertained the nation with anecdotes such as this, his most famous: "[T]he climate is erratic, eccentric and peculiar. The altitude is between 7,000 and 8,000 feet above high water mark, so that during the winter it does not snow much, we being above snow line, but in the summer the snow clouds rise above us, and thus the surprised and indignant agriculturalist is caught in the middle of July with a terrific fall of snow, so he is virtually compelled to wear his snowshoes all through his haying season."[2] Great fun, especially for those of us who wish to preserve low multitudes at high elevations, but simply not true.

Wyoming's Aven Nelson, botanist and tireless advocate for horticulture, observed that it had taken years for permanent residents to come to the realization that, botanically speaking, flowers, forage and forests abound. There are no "deserts" within our borders, he wrote. To early train travelers, it may have looked as though Wyoming consisted of "great barren wastes." But now, with good roads, tourists were coming here "to enjoy with us the charm of the great plains."[3]

In his seminal work, *The Great Plains* (1931), Walter Prescott Webb described the High Plains as constituting the heart of the Great Plains: relatively level, naturally treeless, covered by short-grass, with a semiarid climate, frequented by high winds. Nebraska's Charles Bessey, botanist and dean of Great Plains naturalists, and before him the botanical explorer Joseph Dalton Hooker had used the term "prairie province" to describe the unique geographic distribution of plants in roughly this same region. Most convenient for our purposes, the last superintendent of the Cheyenne Horticultural Field Station defined the High Plains as that area lying west of the 99th meridian (roughly a line from a point fifty miles east of Pierre, South Dakota, through Grand Island, Nebraska, to Hays, Kansas), east of the Rocky Mountains in Colorado and Wyoming, north of the Arkansas River in Colorado and Kansas, and south of the White River in South Dakota—a total of about 120,000 square miles. Average annual rainfall ranges from twelve inches or less on the western side to twenty inches or more on the eastern side. By way of contrast, average annual rainfall is thirty-three inches in Iowa and forty-three inches in Virginia. Elevations on the High Plains range from 1,800 feet on the east to 7,000 feet on the west.[4]

While the word "horticulture" stems from a combination of the classical Latin *hortus* (an enclosure for plants, meaning pleasure garden, fruit garden, kitchen garden, and even vineyard) and *cultura* (meaning to care for or to cultivate), the word was not used before the seventeenth century, and then primarily for fruit- and nut-bearing trees. In 1907, Wyoming's senator Joseph M. Carey, an amateur horticulturist, distinguished between horticulture as the "cultivation of the garden and small field in a great variety of crops, chiefly vegetables, fruits, and flowers, and in an intense way" and agriculture as the "cultivation of the larger fields in less variety of crops, chiefly grasses and grains, and in a wholesale or more comprehensive manner." His friend Aven Nelson later described horticulture as "primarily the growing of garden crops of any kind, including flowers, vegetables, small fruits, standard fruits, ornamentals, and shade trees" for both home and commercial use. Similarly, the enabling legislation for the Cheyenne Horticultural Field Station set forth four major areas for research and experimentation: fruits, vegetables, windbreaks, and ornamental plants—again, a convenient definition for our purposes.[5]

The story of the development of horticulture on the High Plains necessarily pays tribute to the imagination and perseverance of individuals from all walks of life seeking to create livable places out of a vast, seemingly inhospitable piece of space. Chapter 1 describes early horticultural efforts to supplement native and imported foodstuffs with the cultivation of vegetables and fruits and to encourage the planting of trees for protection and flowers for ornaments. Shelter from wind being a prerequisite for horticultural development, Chapter 2 describes early state legislation and citizen activity to promote tree planting. The end of the Civil War and the advent of railroads encouraged the founding of communities on the High Plains, the most notable for horticulture being the Union Colony of Greeley (Chapter 3). In 1878, Major John Wesley Powell's report on the arid West boldly suggested a set of land and water laws and their administration that were entirely different from those that had worked well for the humid East. His report would have far-reaching impact, and thus it serves as a useful foil by which to judge the development of horticulture on the High Plains (Chapter 4).

Passage of the Hatch Act in 1887 provided federal funding to agricultural experiment stations connected to the nation's land-grant colleges, so those colleges became the primary source of knowledge about horticulture. Chapters 5 and 6 describe some of the pioneer horticultural activities of Charles Bessey in Nebraska and Aven Nelson in Wyoming. Ever mindful of their duty to impart what Ben Franklin called "useful knowledge," these pragmatic botanists set the standards as well as the agenda for agricultural experiment stations on the High Plains. Among the pioneer agents of land-grant colleges to the rural population was James E. Payne. His travails on the plains of eastern Colorado serve as moving testimony to the difficulties of introducing horticulture to the most extreme climatic conditions (Chapter 7).

As parts of the High Plains, meanwhile, became more permanently settled and urbanized, enlightened farmers, teachers, and civic leaders turned their attention to community beautification (Chapter 8). More than from anyone else, introduction of hardy plants to the High Plains benefited from the overseas collecting and highly publicized plant breeding conducted by Niels Hansen of South Dakota (Chapter 9).

While horticulture developed most intensively along the Front Range of Colorado, areas more distant from the mountain streams confronted the traditional obstacle of drought and, related to that, the very survival of rural communities. While passage of the Reclamation Act of 1902 set forth a major new role for the federal government concerning agriculture in the arid West, the Smith-Lever Act of 1914 had a more direct and immediate impact on horticulture by institutionalizing and greatly expanding the extension services of land-grant colleges. Those services now went well beyond teaching residents how to grow fruits, vegetables, ornamentals, and trees. Especially after President Theodore Roosevelt's Country Life Commission, federal engagement would encompass literally all those aspects of rural living that made for an improved quality of life (Chapter 10).

Still criticized, both on principle and in practice, the fact remains that the federal government has been indispensable for taming the arid West, just as Major Powell had conjectured. It is within this context, and as the result of one man's passion for community beautification combined with his perfect political connection, that the Cheyenne Horticultural Field Station came into existence. From the early 1930s until the mid-1960s, station staff conducted research on fruits, vegetables, windbreaks, and ornamental plants for which it maintained hundreds of cooperative arrangements with farmers, ranchers, and communities throughout the High Plains (Chapter 11).

The Cheyenne Station's abandonment of horticulture in favor of rangeland research not only signified the political clout of stock farmers but also acknowledged the greatly expanded exploitation of vast sources of groundwater throughout the High Plains. Additionally, technological advances, especially in transportation, meant that residents of the High Plains no longer depended upon themselves for fruits and vegetables.

Both the population explosion along the Front Range and the population decline in smaller communities beyond the Front Range have resulted in renewed interest and activity in horticulture. Ever more mindful of arid conditions, a whole new "green" industry has emerged, and it now contributes greatly to making life throughout the High Plains more pleasant and refined—that is, more civilized (Chapter 12).

We say this despite the fact that qualitative comparisons such as Jefferson made between savage and tame, barbaric and civilized, are now considered invidious. We would suggest, furthermore, that the progress of horticulture in and around especially the more isolated communities of the High Plains has contributed to making life far more pleasant than the much cherished notion of a vast open space where cowboys and other rugged nomads roam without a sense of place.

Notes

1. Pike quoted in Dorothy Weyer Creigh, *Nebraska, a Bicentennial History* (New York: W. W. Norton, 1977), 4–5; Long and James quoted in Maxine Benson, ed., *From Pittsburgh to the Rocky Mountains. Major Stephen Long's Expedition 1819–1820* (Golden, Colo.: Fulcrum, 1988), xiv; James Edwin Bryant, *What I Saw in California, Containing the Complete Original Narrative and Appendix from the 1849 Appleton Edition in True Facsimile* (Palo Alto: Lewis Osborne, 1967), 98; Francis Parkman, *The Oregon Trail* (1849; reprint, New York: Literary Classics of the United States, 1991), 64.

2. T. A. Larson, ed., *Bill Nye's Wyoming Humor* (Lincoln: University of Nebraska Press, 1968), 26–27.

3. Aven Nelson, "The Flora of Wyoming," n.d., box 11, folder 5, Aven Nelson Papers, University of Wyoming American Heritage Center, Laramie.

4. Walter Prescott Webb, *The Great Plains* (Lincoln: University of Nebraska Press, 1981 [1931]), 4, 21, 28; Richard A. Overfield, *Science with Practice: Charles E. Bessey and the Maturing of American Botany* (Ames: Iowa State University Press, 1993), 138; Roger L. Williams, *A Region of Astonishing Beauty: The Botanical Exploration of the Rocky Mountains* (Lanham, Md.: Roberts Rinehart, 2003), 97; Gene S. Howard, "Recommended Horticultural Plants Generally Hardy and Adaptable in the Central Great Plains Region," USDA Agricultural Research Service B-770 (February 1982; reprint September 1999), 6 pp.

5. Paul H. Johnstone, "In Praise of Husbandry," *Agricultural History* 11 (April 1937): 87; Joseph M. Carey, "The Future of Horticulture in the State of Wyoming," *Wyoming State Board of Horticulture Special Bulletin* 1 (1907): 19; Aven Nelson, "Horticultural Department," *Wyoming Farm Bulletin* 5, no. 7 (January 1916): no page.

1

Horticultural Beginnings

Vegetable gardens and ornamental flowers provide the setting for some of the most poignant episodes in Willa Cather's *O Pioneers!* Although fictional, they may well be the most widely read depiction of early settler life on the High Plains. Take, for example, John Bergson addressing his children from his deathbed: "[D]on't grudge your mother a little time for plowing her garden and setting out fruit trees, even if it comes in a busy season. She has been a good mother to you, and she has always missed the old country." Her garden helped Mrs. Bergson reconstruct her former life insofar as possible.

Then, on a September afternoon two years later, Alexandra, the eldest of the Bergson children and by then fourteen, is found by her boyfriend, Carl Linstrum, in her mother's garden, resting from digging sweet potatoes: "[T]he dry garden patch smelled of drying vines and was strewn with yellow seed-cucumbers and pumpkins and citrons. At one end, next to the rhubarb, grew feathery asparagus, with red berries. Down the middle of the garden was a row of gooseberry and currant bushes. A few tough zinnias and marigolds and a row of scarlet sage bore witness to the buckets of water that Mrs. Bergson had carried there after sundown, against the prohibition of her sons."[1]

Willa Cather's contemporary, Charles S. Harrison, a Congregational minister and amateur horticulturist who once lived just a few miles west of the novelist's hometown of Red Cloud, observed: "Many a poor woman on the frontier has slowly faded away with soul starvation. She had potatoes enough, but she needed flowers."[2] Even under the most primitive conditions of early settlement, vegetables fed the body and flowers fed the soul.

On the High Plains, the actual origins of horticulture, in the broadest sense of cultivation of the soil, remain obscure. Some archaeological evidence suggests that prehistoric Plains tribes cultivated the sunflower (*Helianthus annus* L.), but no such evidence is specifically known for the High Plains. Spanish explorers, roaming through the region from Central America, apparently introduced maize, beans, and squashes. Early-nineteenth-century explorers, traders, and trappers occasionally found those plants cultivated around Indian habitations.[3]

The Plains tribes, as we know, were primarily hunters, but they did gather, cook, dry, and process a wide variety of native plants. Among the most common was the prairie turnip (*Psoralea esculenta* Pursh). As late as 1905, Niels Hansen observed tribal members in southwestern South Dakota using these plants: "[T]he Indians dig them out from the prairie sod with a pointed stick and braid them into long chains. When ready to use them, the outer dark brown or blackish coating is removed, leaving the snow white starchy bulb."[4]

Actually, the prairie turnip is not a turnip at all but a legume variously known as scurfy pea, breadroot, Indian breadroot, Indian turnip, prairie potato, pomme blanche, ground apple, white apple, Tipsin, Tipsinna, and Dakota turnip—all of which suggests the wisdom of using the scientific names of plants as well as their horticultural or common names.[5] The scientific names follow *International Rules of Botanical Nomenclature* and thus are universally recognized, but the common names are governed by no formal code and vary from region to region.

At the outset of our story, therefore, and to avoid future confusion, we must understand the rudiments of plant nomenclature. Because Latin was the first, universal language of the sciences, eighteenth-century botanists adapted, and in some cases invented, Latinate words to identify plants, their relationships to each other, and the authors who first described them. Hierarchically from the most general to the most specific, botanists classify plants at six levels: division or phylum, class, order, family, genus, and species. We need

Prairie turnip, Rocky Mountain Herbarium, University of Wyoming, Laramie. Courtesy, Ronald K. Hansen.

concern ourselves only with the last three levels. The family name of a plant is generally recognized by the ending *aceae*; for example, the western sand cherry belongs to the rose family known as *Rosaceae*. Within that family, the western sand cherry belongs to the genus *Prunus*, the genus name always given as a Latin noun. Within that genus, the species name is written as a Latin adjective, in this case, *besseyi*.

It turns out that Liberty Hyde Bailey, dean of American horticulturists and longtime professor at Cornell University, first described the western sand cherry as a separate species in 1898. He named it *Prunus besseyi* in honor of his friend and colleague Charles Bessey of Nebraska. Thus the full scientific name of the western sand cherry became *Prunus besseyi* L.H. Bailey.

To somewhat complicate the matter of nomenclature, especially for those of us with little or no background in the sciences, the systematic classification of plants is fluid rather than static, changing as a result of new research and other factors. To continue our example, Henry Allan Gleason (Gl.) of the New York Botanical Garden reclassified the western sand cherry in 1952, from a separate species to a variety or subspecies of the sand cherry (*Prunus pumila* L.)—the latter first described by Carl Linnaeus (L.), the founder of modern taxonomy. As a result, the western sand cherry is now known and written as *Prunus pumila* L. var. *besseyi* (Bailey) Gl.[6]

In addition to the plants created in nature, a great number of varieties have been developed through plant propagation and plant breeding. Known as cultivars, the names of these varieties are generally given in English and written in single quotes, such as *Fragaria vesca* L. cv. 'Ogallala' for the strawberry cultivar developed at the Cheyenne Field Station from the crossing of a hardy native plant with a large commercial variety.

Because the early traders were essentially hunters, the numerous edible plants native to the High Plains undoubtedly played an insignificant role in relieving starvation. With the establishment of trading outposts on the High Plains in the 1820s and 1830s, efforts certainly were made, albeit isolated, to grow vegetables—for example, at Bent's Fort on the Arkansas River, Lupton's Fort on the South Platte, and Fort William (later renamed Fort Laramie) on the North Platte. If the recollections of Benjamin Louis Eulalie de Bonneville (1832) are any indication, horticulture at Fort William generally did only marginally well: "All attempts at agriculture and gardening in the neighborhood ... have been attended with very little success. The grain and vegetables raised there have been scanty in quantity and poor in quality." Given the aridity and the elevation (4,300 feet), the region was slated to remain forever in "a state of pristine wildness."[7]

That view was confirmed by Edwin Bryant, a Kentucky journalist who stopped at Fort William fourteen years later, in June 1846, on his way to California. "Not a foot of ground around the fort is under cultivation," he reported. "Experiments have been made with corn, wheat and potatoes, but they either have resulted in entire failures, or were not so successful as to authorize a renewal." In addition to the adverse climatic conditions, Bryant

Soldiers protecting vegetable garden, Fort Laramie, ca. 1880. Courtesy, Fort Laramie National Historic Site.

suggested another reason for crop failures: "The Indians, who claim the soil as their property, and regard the Fur Company as occupants by sufferance, are adverse to all agricultural experiments; and on one or two occasions they entered the small enclosures, and destroyed the young corn and other vegetables as soon as they made their appearance above the ground."[8] After the U.S. government purchased the fort in 1849, the military at certain times of the year posted guards around the clock to protect its gardens.

Indeed, since 1818 the War Department had specified that soldiers at every military post "will annually cultivate a garden . . . equal to supplying hospital and garrisons with the necessary kitchen vegetables throughout the year" and that the commanding officer "will be held accountable for any deficiency in the cultivation."[9] That was a tough order for any post on the High Plains, although surprisingly well accomplished at Fort Laramie beginning with the 1850 growing season, flourishing after the Civil War, and continuing until the fort's abandonment in 1890.

The War Department's order of September 11, 1818, had been given for reasons of both health and finance. By then, it was well understood that vegetables and fruits were essential to good health, most specifically for preventing the debilitating effects of scurvy, which we now know are a result of diets deficient in Vitamin C. In addition, the War Department sought to limit transportation expenses by having soldiers, so far as possible, grow their own produce. The high cost of transporting bulk goods, before the advent of the railroad, would serve both as obstacle to importing plant materials and as incentive for local horticulture.

Fort Laramie was among the outposts most distant from supply depots, strategically situated 600 miles west-northwest of Fort Leavenworth (the beginning of the Oregon Trail), at the confluence of the Laramie and North Platte rivers on the High Plains of eastern Wyoming. In the early spring of 1850, Fort Laramie's supply officer recorded that he had gotten ten acres "sod-busted," that he had secured plows to cultivate the land, and that he would "put in as much seed, corn, oats and barley, as my means will allow." He had also secured the services of a settler, recently arrived from the Arkansas River valley and knowledgeable about irrigation. That arrangement did not last long, as the settler joined a convoy en route to the gold mines of California. This led the officer to send an agent to Taos to recruit ten to twelve Mexicans, knowing they would be familiar with irrigated farming, which the supply officer believed was the only way to successfully cultivate around Fort Laramie. The officer also noted that Mexicans worked more cheaply than Americans.[10]

The earliest irrigation at Fort Laramie consisted of a single earthen ditch, no more than a few hundred yards in length, taking water out of the Laramie River for gardens on nearby bottomland. Nothing indicates that seeds were purchased from nearby tribal members or settlers or that the soldiers had collected seeds of native edible plants. Instead, the post supply officer requisitioned seeds from Fort Leavenworth. In late summer of 1856, he reported that, while his potato plants looked well, the fact that his seed potatoes had arrived in poor condition meant the harvest "will be very small, if it does not fail entirely." As a result, he requested his counterpart at Fort Leavenworth to take great care not only in selecting the best plants but also in carefully packing them in hay to protect against freezing while on the trail.[11] The extraordinary difficulties of transporting plant materials long distances would remain a major challenge to the development of horticulture on the High Plains.

That fact makes all the more remarkable the stories of the earliest settler women who brought to Nebraska cuttings of their favorite houseplants—geraniums, for example—as reminders of the civilization they had left behind. To be sure, they also collected native plants, providing both food to the table and pleasantness to the home. Along watercourses they could find several species of greens such as lamb's quarter (*Chenopodium berlandieri* Moq.), asparagus (*Asparagus officinalis* L.), and onion (*Allium canadensis* L); on hillsides and prairie ravines they could find the American plum (*Prunus americana* Marsh.), chokecherry (*Prunus virginiana* L.), red currant (*Ribes cereum* Dougl.), black currant (*Ribes americanum* P. Mill.), buffalo currant (*Ribes odoratum* Wendl.), and bush grape (*Vitis acerifolia* Raf.). In sandy, rocky areas they could find buffaloberry (*Shepherdia argentea* [Pursh] Nutt.) and the sand cherry (*Prunus pumila* L. var. *besseyi*). Their selection of ornamentals for transplanting likely converged on those plants found similar to the plants they knew back East. In addition, we note early descriptions of dugouts and sod houses, both with wildflowers growing on their roofs.[12]

Since the time of the earliest homesteads, horticulture has served a distinctly palliative role in making life more pleasant for women. Unquestionably, loneliness, insecurity, and hard conditions on the High Plains affected women more than men. Testimonials to the salutary effects of horticulture over the harshness of life continued throughout the history of the High Plains. In the thrilling story based on the life of Jules Sandoz, a neighbor returns to his western Nebraska homestead on a cold January day, having been away to earn enough money to support his family, and finds his wife and three children dead. Later, a neighbor woman, helping to prepare for the funeral, reflects sorrowfully on the deceased wife: "If she could [have] had even a geranium—but in that cold shell of a shack" that was not possible.[13]

If one had to choose a single activity to represent the advent of permanent human settlement on the High Plains, it would be tree planting. When settlers began arriving, "[T]rees were so rare in that country," observed Jim Burden, narrator of Willa Cather's *My Antonia*, "and they had to make such a hard fight to grow, that we used to feel anxious about them, and visit them as if they were persons."[14]

Tree planting began in southeastern Nebraska during the 1850s and spread west beyond the 98th meridian by 1860. The earliest tree planters took shade and forest tree cuttings from along watercourses and dug up saplings for transplanting. The earliest fruit trees, however, came from points east of the Missouri River. In 1856, J. Sterling Morton, who would found Arbor Day and later became secretary of agriculture under President Grover Cleveland, imported 500 apple trees for his Nebraska City farm; and Robert W. Furnas, Nebraska's second governor, established the state's first commercial nursery, at Brownville, also in the mid-1850s.[15]

Though partisan political adversaries, Morton and Furnas helped pass the earliest state legislation pertaining to horticulture on the High Plains. By act of the Nebraska Territorial Legislature in 1861, any Nebraska property owner who planted at least 100 fruit or ornamental trees, or 400 forest trees, per acre received a fifty dollar exemption on the valuation of that property. This act proved so popular that the resulting decline in tax revenues drove the legislature to repeal it in 1864.[16]

While Nebraskans, from east to west, deservedly earned the reputation of tree planters, Coloradoans along the Front Range successfully established themselves as market gardeners and orchardists. The impetus for such horticulture came as a result of the demand for foodstuffs generated by the gold and silver rush into the Rocky Mountains; its development depended on successfully harnessing the water flowing out of the mountains onto the plains. For those who did not see the Front Range before the 1960s, it may be impossible to imagine that at one time, dotted with gardens, farms, and orchards, this was one of the world's great intensive-agriculture regions.

David K. Wall, who arrived in 1859 from Indiana via California, where he had raised and supplied food for miners, is considered the first market gardener to use irrigation in Colorado. He took enough water out of Clear Creek to irrigate about two acres near Golden. Concerning such early irrigation efforts, Elwood Mead, a pioneer irrigation engineer, observed that generally brush and stones were used to deflect streams. Irrigators probably just used a few shovelfuls of earth to make embankments and then shoveled open passages to cultivated plots as water was needed. Wooden and iron headgates to control water flow into ditches came later. By 1862, three years into the Rocky Mountain gold rush, waters for irrigation were being taken out of all the main streams within the upper South Platte watershed: Clear Creek, Boulder Creek, and the St. Vrain, Big Thompson, and Cache la Poudre rivers.[17] Large-scale irrigation projects along the Front Range, however, would not begin until the 1870s, after the passage of legislation regulating water and the creation of irrigation institutions—partnerships, community cooperatives, corporations, and districts.

Market gardening, meanwhile, gained an early publicist in William N. Byers (1831–1903), founding editor of Denver's *Rocky Mountain News*. An outspoken advocate of agricultural and horticultural development, Byers believed that without such production, Denver could never become a major city. On May 7, 1859, during the newspaper's first spring of publication, Byers reported that David Wall of Golden had "left at our office a large supply of garden seed for sale. All ye that wish fresh vegetables walk up and select your packages at 25 cents each." On June 11, Byers acknowledged receiving radishes, which he believed were the first ever grown in Colorado; and on June 25 he announced that locally grown peas, lettuce, and onions were available for purchase. On August 13 Byers reported that "our market is now well supplied with garden vegetables of as fine quality as can be found in the old settlements of the States," which included cabbages, melons, and squashes shipped to Denver from farms in the Arkansas River valley.[18]

Orchards were more difficult to establish than market gardens, in part because of the costs and risks involved in importing the nursery stock by wagon over long distances. In 1862, for example, one Henry Lee brought 6,000 apple cuttings and 500 each of peach, pear, plum, and cherry cuttings from an orchard at Iowa City to his brother's farm near Golden. While we have no record of exactly how Lee transported the cuttings, we do know that fifteen years earlier a pioneer nurseryman named Henderson Luelling left Salem, Iowa, with an ox-drawn wagon packed with 700 fruit cuttings, stopped at Fort Laramie, and reached the Willamette Valley of Oregon after five months on the trail. Approximately half of his cuttings survived the trip, enough for him to start a thriving nursery business that eventually led him to California's Central Valley. Luelling's cuttings had ranged from twenty inches to four feet in height; they were planted in a mixture of soil and charcoal, which held water better than soil alone, in two specially built boxes that took up an entire wagon bed, surrounded for protection by a light but sturdy frame.[19]

As to Lee's stock, we do not know how many trees survived his trip, but we do know that in 1864 he salvaged only 150 after a flood along Clear Creek. He then moved and replanted to nearby upland and by 1866 was again cultivating a thriving orchard. Further north near Bellvue, where the Cache la Poudre River emerges from the mountains, a settler started an orchard in 1862 from imported cuttings, species unknown, transported across the plains; and in 1863 another settler established the first apple orchard in the Big Thompson valley west of present-day Loveland.[20]

Whether market gardeners or orchardists, early growers along the Front Range, generally speaking, were squatters—that is, settlers with no legal title to the land they occupied. To provide themselves with some security against latecomers, speculators, and territorial administrators, these cultivators came together in voluntary, protective associations known as claims clubs. In the case of the Arapahoe County Claims Club, established in 1859, each member publicly stated his claim, giving its geographic description; to confirm the validity of the claim, the member agreed to "make, or cause to be made, improvements on his or their claim, by breaking one acre of land; or building a house sufficiently good to live in."

Because the institution of the claims clubs clearly encouraged horticulture, it is worth reflecting on their nature. We often view the development of institutions in the arid West as somehow entirely different than those in the East, when, in fact, both are based on the liberal and associational principles that grew out of the eighteenth-century Enlightenment in Europe and America. Witness the familiar ring of the preamble to the bylaws of the Arapahoe County Claims Club: "[W]hereas it sometimes becomes necessary for persons to associate themselves together for certain purposes, such as the protection of life and property; and as we have left the peaceful shade—left friends and homes for the purpose of bettering our condition, we, therefore, associate ourselves together."[21]

By 1861, claims clubs covered a considerable portion of the Front Range and had gained recognition by both the U.S. Congress and the Colorado Territorial Legislature. By serving as vehicles for resolving conflicts over land ownership until state and federal laws took hold, these voluntary associations promoted more permanent cultivation and, in time, the development of communities.[22]

Consistent with the purpose of the claims clubs, President Abraham Lincoln submitted, and Congress ratified, the Homestead Act in 1862. As with the bylaws of the claims clubs, the Homestead Act did require certain improvements of the land. Essentially, the act provided that any citizen could obtain title to 160 acres of unappropriated public land by residing on or cultivating that land for a period of five years and paying very modest filing fees.

Much has been written about the Homestead Act. While its purpose was lauded by many, its actual impact on the settlement of the High Plains proved not altogether positive. Most immediately, it was used by land promoters who unscrupulously argued that one could sustain a family and produce surplus food for the market on 160 acres of arid land. Loopholes in the act combined with Congress's inaction led to wild land speculation. The impracticability of the act's acre limit stirred Walter Prescott Webb, a westerner, to express the view that "no law has ever been made by the Federal government that is satisfactorily adapted to the arid region." To this day, Webb's view remains an ever-popular opinion throughout the High Plains.[23]

Were it not for two other major pieces of legislation proposed and signed by President Lincoln, there would be no horticulture as we know it on the High Plains: one creating the United States Department of Agriculture on May 15, 1862, and the other establishing the land-grant colleges, July 2, 1862. In the first, "[T]he general designs and duties of [the department] shall be to acquire and to diffuse among the people of the United States useful information on subjects connected with agriculture in the most general and comprehensive sense of that word, and to procure, propagate, and distribute among the people new and valuable seeds and plants." In the second, also known as the Morrill Act, each state obtained public land and an appropriation for an agricultural college.[24] Taken together, these two acts launched the federal government's support of scientific research and its application to the practice of agriculture and related fields. The long-term effect on the development of horticulture on the High Plains would be incalculable. Meanwhile, state and local governments had taken certain actions that contributed more immediately to the development of horticulture.

Notes

1. Willa Cather, O Pioneers! (1913; repr. Boston: Houghton Mifflin, 1988), 27–28, 48–49.

2. Charles S. Harrison, The Gospel of Beauty and Kindred Topics (York, Neb.: privately printed, 1917), 87.

3. Kelly Kindscher, Edible Wild Plants of the Prairie, an Ethnobotanical Guide (Lawrence: University of Kansas Press, 1987), 127; William H. Alderman, ed., Development of Horticulture on the Northern Great Plains (St. Paul: Great Plains Region, American Society for Horticultural Science, 1962), 27, 120.

4. Niels E. Hansen, "Some Horticultural Questions," South Dakota Horticultural Society Annual Report (1907): 164.

5. Waldo R. Wedel, "Notes on the Prairie Turnip (Psoralea esculenta) among the Plains Indians," Nebraska History 59, no. 2 (Summer 1978): 155.

6. Henry A. Gleason, The New Britton and Brown Illustrated Flora of the Northeastern United States and Adjacent Canada (New York: New York Botanical Garden, 1952), 2, 330.

7. Edgeley W. Todd, ed., The Adventures of Captain Bonneville U.S.A. in the Rocky Mountains and Far West Digested from His Journal by Washington Irving (Norman: University of Oklahoma Press, 1986), 37.

8. James Edwin Bryant, What I Saw in California, Containing the Complete Original Narrative and Appendix from the 1849 Appleton Edition in True Facsimile (Palo Alto: Lewis Osborne, 1967), 109. For an exciting, although politically incorrect, account of life around Fort Laramie, see Francis Parkman, The Oregon Trail (1849; repr. New York: Literary Classics of the United States, 1991).

9. Quoted in Miller J. Stewart, "To Plow, to Sow, to Reap, to Mow: The US Army Agriculture Program," Nebraska History 63 (Summer 1982): 194.

10. Captain Stewart Van Vliet, assistant quartermaster, Fort Laramie, to Major General T. S. Jessup, quartermaster general, Washington, D.C., April 9, 1850, and undated letter, copies, Fort Laramie National Historic Site Library, Fort Laramie, Wyoming (hereafter cited as FL MSS).

11. Major W. Hoffman, Fort Laramie, to Lieut. J. L. Corley, adjutant general, St. Louis, August 10, 1856, FL MSS.

12. Kindscher, *Edible Wild Plants*, 5; Dorothy Weyer Creigh, *Nebraska, a Bicentennial History* (New York: W. W. Norton, 1977), 89–92.

13. Mari Sandoz, *Old Jules, Portrait of a Pioneer* (1935; repr. New York: MJF Books, 1996), 83. For a portrayal of the "spiritual effect" of the Plains on women, see Webb, *The Great Plains*, 505–506.

14. Willa Cather, *My Antonia* (1918; repr. Boston: Houghton Mifflin, 1977), 29; see also Kathleen Norris, *The Cloister Walk* (New York: Riverhead Books, 1996), 287–296.

15. Nebraska State Horticultural Society Proceedings (1872): 11.

16. Everett N. Dick, *Conquering the Great American Desert: Nebraska* (Lincoln: Nebraska State Historical Society, 1975), 85, 117–118.

17. Alvin T. Steinel, *History of Agriculture in Colorado* (Fort Collins: Colorado State Agricultural College, 1926), 180; Robert G. Dunbar, *Forging New Rights in Western Waters* (Lincoln: University of Nebraska Press, 1983), 122.

18. Steinel, *History of Agriculture*, 180–183; Byers quoted in Leroy R. Hafen, ed., *Colorado and Its People* (New York: Lewis Historical Pub., 1948), 1:182 (first two quotes), 1:183 (last quote).

19. J. S. Stranger, "Reminiscences of Orcharding in Colorado," Colorado State Board of Horticulture Annual Report 14 (1902): 101; Thomas C. McClintock, "Henderson Luelling, Seth Luelling and the Birth of the Pacific Coast Fruit Industry," *Oregon Historical Quarterly* 60, no. 2 (June 1967): 156–157.

20. Ansel Watrous, *History of Larimer County, Colorado* (Fort Collins: Courier, 1911), 143; Winona W. Taylor, "Progress of Horticulture in the Big Thompson Valley," Colorado State Board of Horticulture Annual Report 1 (1889): 293.

21. Quoted in Hafen, *Colorado and Its People*, 1:214.

22. Steinel, *History of Agriculture*, 43–45.

23. Webb, *The Great Plains*, 398–399.

24. Gladys L. Baker and others, *Century of Service: The First 100 Years of the United States Department of Agriculture* (Washington, D.C.: United States Department of Agriculture, 1963), 13.

2

Trees for the High Plains

Nebraskans take their trees very seriously. By an act of 1873, the state legislature provided that any person who willfully and maliciously injured or destroyed any trees, valued at thirty-five dollars or more, on the property of another was subject to imprisonment in the penitentiary, hard labor for no less than one year or more than ten years, and liable for double damages to the injured party. In Nebraska, state historian Everett Dick observed, one could plead self-defense for shooting someone and get away with it, "but no such subterfuge could be claimed by a tree mutilator."[1]

Until 1945, Nebraska's official nickname was "Tree Planters' State," in recognition of tree-planting activities such as the institution of Arbor Day, founded in 1872, inspired by J. Sterling Morton; passage of the Timber Culture Act in 1873, sponsored by U.S. Senator Phineas W. Hitchcock of Nebraska; and establishment of the first and only artificial national forest near Halsey in 1902, promoted by Professor Charles Bessey. The legislature later changed Nebraska's nickname to "Cornhusker State" in recognition of the

University of Nebraska football team, which by 1945 had changed its name from "Bugeaters" to "Cornhuskers."

Long before football became Nebraska's civic religion, the state legislature had promoted the planting of trees for purposes of windbreaks, orchards, woodlots, and beautification. By an act of 1869, the legislature confirmed property tax exemption for tree planting, allowing up to $100 exemption annually for five years to citizens planting one or more acres with woodlot trees no more than twelve feet apart and providing an additional $50 exemption for every acre planted in fruit trees. Two years later the legislature encouraged private owners of town lots to plant trees along public rights-of-way by levying a special tax of $1 to $5 per year for each lot adjacent to a street not planted. The goal was to line all Nebraska streets with shade trees within four years. Furthermore, the 1871 legislature made stockgrowers liable for damages caused by their herds on cultivated lands, broadly defined to include forest trees, fruit trees, and hedgerows.[2] Therein was fired an early volley in the contest between nomadic cattlemen and permanent settlers in western Nebraska.

Initiative for tree legislation and promotion of tree planting came from the Nebraska State Board of Agriculture and, in particular, from the State Horticultural Society, which grew out of the former. During the 1869 state fair, "friends of horticulture" gathered at the Board of Agriculture office to establish the Nebraska State Horticultural Society. The society was formally incorporated and recognized in 1871 by the legislature, with an appropriation of $2,000 to the State Board of Agriculture designated for support of the society's activities. Robert Furnas and J. Sterling Morton were among the twenty-three founding members. Its overall purpose: "the promotion of Pomology, Arboriculture, Floriculture, and Gardening."[3]

At the beginning, society activities centered first on determining by trial and error which fruit varieties purchased in nurseries east of the Missouri River did best in Nebraska and then on displaying those successes both within and outside the state. In 1873, for example, Morton used his position as a real estate promoter for the Burlington and Missouri Railroad to convince the railroad to provide free transportation in a specially built car for Nebraska-grown fruit to an exposition in Boston.[4]

While the earliest fruit-growing successes—apples, pears, peaches, grapes, plums—occurred in southeastern Nebraska and not on the High Plains, the membership of the society did include individuals interested in understanding the horticultural possibilities of the High Plains. Preeminent among them was Charles S. Harrison (1832–1919), a preacher-horticulturist who had emigrated from north-central Illinois to York, about fifty miles west of Lincoln,

Family with newly planted trees, near New Helena, Custer County, Nebraska. Courtesy, Nebraska State Historical Society.

between the 97th and 98th meridians. There, he created his own horticultural experiment farm, and for nearly a half century he crisscrossed the High Plains as a speaker on horticultural topics and served as a frequent contributor to *Field and Farm*, a Denver-based newspaper—always as a missionary for horticulture in the arid West.

In 1872, Harrison won first prize in the inaugural essay competition sponsored by the Nebraska State Horticultural Society. His "Essay on Tree Culture" addressed the question of why Nebraskans should plant more trees. His answers, he said, were based on what he had observed on trips through Michigan, Wisconsin, and Minnesota. He was neither the first nor the last person who sought to apply knowledge of the humid, forested states to the arid, treeless states. His thesis was that as rainfall had followed tree belts and waterways in the Midwest, so too planting trees on the Nebraska plains would produce more rainfall and thus allow for more horticulture and more permanent settlement.

Harrison's essay, furthermore, conveyed a sense of the inevitability of harnessing nature for the benefit of humanity. "Providence," he wrote, "seems to encourage the adventure of men as they push Westward. The mythical desert will doubtless be covered with beautiful groves and fruitful orchards even to the base of the Rocky Mountains." He quoted Scripture (Isaiah 41:18–19) to support that sense of Manifest Destiny: "I will open rivers in high places, and fountains in the midst of the valleys; I will make the wilderness a pool of water, and the dry land springs of water. . . . I will set in the desert the fir tree, and the pine, and the box tree together."[5]

In a more secular vein, Harrison cited the example of the landmark French forest legislation of 1860, which combined government subsidies, tax exemptions, and technical advice to landowners. Encouraged by this and other European precedents, William Stolley of Hall County, farming west of the 98th meridian near Grand Island and a member of the State Horticultural Society, proposed that the legislature enact a law to ensure that there would be at least one 160-acre grove in each of Nebraska's ninety-plus counties.[6]

As in Nebraska, state horticultural societies provided an institutional framework for the promotion of tree culture in Kansas (established 1867), Colorado (1880), South Dakota (1890), and Wyoming (1907). Overlooked is the fact that in purpose, outlook, and activities, these societies closely resembled the agricultural societies of Europe and America in the eighteenth century. Shared in common was the desire to spread the latest scientific knowledge among all cultivators, although one is left with the impression that members communicated mostly among themselves and not with those most in need of enlightenment. Paid memberships remained relatively small, for example, fewer than fifty in Nebraska in the 1880s and fewer than a dozen in Wyoming in the 1910s. The state societies generally assembled in their respective conventions annually or semiannually. State legislative subventions, although consistently meager, made possible publication of their proceedings, which included formal papers, transcripts of discussions, lists of plants adapted to specific geographic areas, and announcements of prizes for the winning horticultural exhibits at county and state fairs. These publications reveal the thinking and practices of the most progressive horticulturists on the High Plains from the late 1860s through the 1920s.

At its first meeting following official recognition by the legislature in 1871, the Nebraska State Horticultural Society entrusted J. Sterling Morton, editor of the Nebraska City *News*, "to prepare and publish an address to the people of the State setting forth all important facts relative to fruit growing in Nebraska."[7] His "Fruit Address," given at the society meeting in Lincoln on January 4, 1872, set forth not simply the "facts" but also the civilizing rationale for such planting.

As perhaps the earliest full exposition of a rationale that would be repeated time and again, Morton's words merit a close review:

There is beauty in a well-ordered orchard which is a joy forever. It is a blessing to him who plants it, and it perpetuates his name and memory, keeping it fresh as the fruit it bears long after he has ceased to live. There is comfort in a good orchard, in that it makes the new home more like the old home in the east, and with its thrifty growth and luscious fruits, sows contentment in the mind of a family as the clouds scatter the rain. Orchards are missionaries of culture and refinement. They make the people among whom they grow a better and more thoughtful people. If every farmer in Nebraska will plant out and cultivate an orchard and a flower garden, together with a few forest trees, this will become mentally and morally the best agricultural state, the grandest community of producers in the American union. Children reared among trees and flowers growing up with them will be better in mind and in heart, than children reared among hogs and cattle. The occupations and surroundings of boys and girls make them, to a great extent, either bad and coarse, or good and gentle. If I had the power I would compel every man in the state who had a home of his own, to plant out and cultivate fruit trees.[8]

While the "Fruit Address" served as the platform, Morton's resolution to create and name Arbor Day was the clarion call for a "campaign for tree planting." That resolution, introduced to the State Board of Agriculture also on January 4, 1872, and unanimously adopted, provided for the board to award $100 to the county agricultural society that on that designated day "planted properly" the largest number of trees and to present a library valued at $25 to the individual who on that same day planted the greatest number of trees. The idea, Morton reminisced thirty years later, "was to concentrate all the thought of the commonwealth on a single day, upon the very important topic of tree planting."[9]

Although the Board of Agriculture had previously offered some prizes for tree planting, Morton's resolution had the effect of exciting general public interest, with over 3 million trees planted in Nebraska on the first Arbor Day, celebrated April 10, 1872.[10] Thereafter, the governor annually proclaimed Arbor Day until 1885, when the legislature permanently designated April 22, Morton's birthday, as Arbor Day.

In addition to leading voluntary and local support for tree planting, members of the Nebraska State Board of Agriculture and the Nebraska State Board of Horticulture successfully lobbied the U.S. Congress and President Ulysses S. Grant for federal aid to tree planters. Introduced by Senator Hitchcock of Nebraska, the Timber Culture Act of 1873 provided 160 acres of unappropriated public prairie land to anyone who would agree to plant trees on forty of those acres, not more than twelve feet apart, and to keep them healthy for eight years. From the beginning, however, there were problems with the act, causing numerous amendments—all unsuccessful—to reconcile its intent with arid conditions. The underlying obstacle, Walter Prescott Webb noted in retrospect, was the act itself: "Not only was it impossible to legislate forests onto the Plains, but it was impossible to make them grow in the arid portions."[11] Before the act's repeal in 1891, omissions in the law had resulted in numerous abuses detrimental to the development of horticulture. In western Nebraska, for example, the lack of residence requirement allowed speculators, cattlemen in particular, to amass new claims beyond the acre limits of the Homestead Act and earlier preemption acts.[12]

An entirely different approach to making life more comfortable on the High Plains came from Edwin A. Curley, an English gentleman farmer who toured Nebraska in 1875 and then, as an enthusiast for settlement, wrote a practical guide for prospective immigrants. On his own estate, Curley had used hedges—closely planted lines of bushes, small trees, or dead wood—as boundaries and means of protection for fields, gardens, and roads. Based on his travels and through his guide, he is credited with introducing to Nebraskans the idea of using hedges as a means of protection from the wind.

Curley reported that the most popular hedge in Nebraska was the Osage-orange (*Maclura pomifera* [Raf.] Schneid.). A member of the mulberry (Moraceae) family, the Osage-orange is so named because of its rough-textured fruit that, from a distance, looks somewhat like green oranges. A dense, spiny, and hardy small tree, the Osage-orange is native to Arkansas, Oklahoma, and Texas. After its introduction to the eastern edge of the High Plains, the Osage-orange spread prolifically without cultivation, very much like the Russian-olive (*Elaeagnus angustifolia* L.) of later popularity. Both trees are considered adventives, that is, not native but fully adapted to their new surroundings. Popularly described by farmers as "horse high, bull strong, and hog tight," the Osage-orange functioned as the most common fencing on the prairie. After the introduction of barbed wire in the mid-1860s, however, the Osage-orange was dug up as a nuisance plant.

Actually, Curley preferred willows (*Salix* L.) as a more aesthetically uniform hedge material, although he recognized that willows, when still young, require protection from the elements and livestock. Whether for use as hedges or in woodlots, Curley came up with the idea that trees and shrubs "be planted in belts, transverse to the prevailing direction of the wind." Such planting, he believed, would protect the immediate surroundings of the homestead from the wind and slow down the prevailing wind speeds. He argued that if a cold north wind or a hot south wind could be substantially slowed from thirty to

Farmstead, Custer County, Nebraska, 1888. Courtesy, Nebraska State Historical Society.

perhaps ten miles per hour, horticulture would benefit greatly. Because planting in belts needed to occur on a large scale to be effective, Curley insisted that state legislatures provide both moral and financial support. While the relationship is unclear, the Nebraska Legislature in 1879 did authorize county commissioners to pay $3.88 per acre out of local funds for the establishment of windbreaks. The legislature specified that to qualify for such support, a "belt" had to consist of at least six rows of trees eight feet apart and trees within rows planted no more than four feet apart and had to be properly maintained for five successive years.[13]

As in Nebraska, tree planting in Kansas began in the east, then moved west across the High Plains. In the early days, when Fort Leavenworth still served as supply depot for Fort Laramie, settlers around that eastern Kansas outpost cultivated several unspecified varieties of apples that had originally come from Russia.[14] In 1865, four years after statehood, the Kansas Legislature approved the first in a series of laws to encourage tree planting. Counties, for example, could pay fifty cents per acre annually to anyone who planted five acres or more of trees, starting two years after the trees were planted and lasting for a period of twenty-five years. The premium for individual settlers increased to two dollars per acre in 1868, and the law was broadened to enable counties to pay two dollars per half mile of trees planted along public thoroughfares, provided the trees were placed no more than one rod (16.5 feet) apart and maintained for three years. For reasons unknown, but possibly because of counties' unwillingness to contribute monies, the legislature repealed this law in 1874.[15]

Meanwhile, the *Kansas Farmer*, established in 1863 as the earliest agricultural newspaper covering the High Plains, spearheaded the movement to establish a Kansas state horticultural society. The immediate consideration was to organize against unscrupulous tree vendors who were "recommending and selling . . . varieties of fruit trees and plants that are entirely unsuited to our climate, as well as under erroneous names." As long as they published, *Kansas Farmer* and, after 1886, *Field and Farm* frequently reported complaints against tree peddlers while, at the same time, the newspapers vigorously editorialized for the establishment of more local tree nurseries.

Kansas Farmer utilized its mailing list to attract fruit growers, gardeners, nurserymen, and amateur horticulturists to join together in the first state horticultural society to include a section of the High Plains, established at Leavenworth on December 10, 1867. Reflecting the emphasis on fruit culture, it was initially known as the Kansas State Pomological Society.[16]

While fruit culture made little immediate headway on the High Plains of western Kansas, the planting of shade trees and windbreaks began in the early 1870s. Notorious in his support of tree planting, Richard S. Elliott was one of the early "boomers." As a real estate promoter (his official title, "Industrial Agent") for the Kansas Pacific Railroad, in 1870 he persuaded the company to underwrite the development of experimental tree plots near the railway stations of three western Kansas communities: Wilson (east of Hays), Ellis (west of Hays), and Wallace (near the Colorado state line). Elliott sought to prove that trees could grow without irrigation on the treeless plains, that more trees would attract more rainfall, and that increased rainfall would greatly enhance opportunities for farming in the arid West as had occurred in the humid East—all destined to improve business for the Kansas Pacific Railroad. As Samuel Aughey and Charles Dana Wilber later did in Nebraska, Elliott combined what we would now call pseudo-science with sheer force of conviction; in his case, every indication suggested that he truly believed what he proclaimed.

In advance of the 1871 growing season, Elliott announced to readers of *Kansas Farmer* that the Kansas Pacific Railroad would be giving away tree seedlings, that he expected 50,000–100,000 to be planted that first year, and if they did well, he foresaw that "millions of trees will soon be growing in Central and Western Kansas." In a subsequent letter to the editor of *Kansas*

Farmer, Elliott strongly urged farmers in western Kansas to begin collecting, planting, and cultivating seeds of native trees. He singled out the white ash (probably *Fraxinus americana* L.), the boxelder (*Acer negundo* L.), and the cottonwood (probably *Populus deltoides* Marsh.).[17]

As one of his promotional activities, Elliott conducted train excursions for journalists, representatives of the federal government, and influential citizens from eastern Kansas and Missouri to visit areas newly opened to settlement in western Kansas. In June 1871, for example, he took members of the Kansas State Agricultural Society to visit his three experimental tree plots. At Wallace, Elliott told journalist John Tice that various deciduous trees had been planted from seed, and Tice reported that they were flourishing.[18] To be sure, May and June naturally tend to be the greenest months on the High Plains. Two years later, drought combined with financial panic ended both Elliott's tree experiments and his career with the Kansas Pacific Railroad.

Several years later, Martin Allen of Hays recalled his first visit to the railroad grounds at Ellis in 1872, where he had seen two acres of forest trees set in nursery rows. On a return visit in 1878, few, if any, of those trees had survived. Much like Nebraska's Charles Harrison, Allen both preached and practiced adapting horticulture to the High Plains. He too, had emigrated from Illinois. In 1873 he had shipped cuttings of trees and shrubs from Illinois to his new homestead. Drought and grasshoppers, however, destroyed all the saplings as well as plants generally considered indestructible, such as asparagus, horseradish, and rhubarb. Allen put some of the blame for that destruction on himself, admitting that he had not been "on the ground in person" to make sure his new land was properly cultivated and his cuttings were of top quality.

Despite setbacks, Allen kept trying. In 1875 he procured new trees and vines from Lawrence. The former arrived infected with insect borers, and the vines came already dead. Nonetheless, by 1878, he reported from Hays that he had successfully harvested cultivar gooseberries, strawberries, raspberries, grapes, and peaches; he anticipated harvesting apples, pears, and plums. He had also gathered wild plums and grapes (*Prunus gracilis* Engelm. & Gray and *Vitis acerifolia* Raf. are native to the Hays area) but said nothing about preserving their seeds and attempting to cultivate them.

For protection from wind and for fuel, Allen had taken advantage of the Timber Culture Act by planting thousands of year-old cottonwood seedlings (varieties not specified) on his own claim. Altogether, more than 2 million acres of Kansas prairie would be claimed under this act, although by the 1930s less than 5 percent of that acreage was forested. Allen also had experimented with cultivating native trees—ash, boxelder, honey-locust, mulberry, hackberry, red elm, and willows—and he noted that much more work had to be done to determine the best species and varieties for local soil and climatic conditions.

Allen took special interest in the matter of hardiness under conditions of summer heat, wind, and dryness. That led him, in 1878, to propose the establishment of a state-supported horticultural school where "men of science" could test plants under field conditions. With all existing state institutions located in the eastern third of the state, Allen argued that it was time for the legislature to "let us have a new and useful one in the west."[19] Although never a horticultural school, the Hays substation of the Kansas Agricultural Experiment Station would provide the setting to test plants under local conditions beginning in 1901.

At the western edge of the High Plains, meanwhile, a few intrepid amateur horticulturists, benefiting from isolated trials and errors before and during the Civil War, developed a veritable network of commercial orchards along the Front Range, from the foothills along the Arkansas River valley near Cañon City to the Cache la Poudre River valley near Fort Collins. Precipitation along the Front Range does not differ appreciably from that on the western Kansas prairie, but early Colorado orchardists had discovered that although fruit trees required little irrigation, that which was needed could be obtained with relatively little effort by drawing from the streams flowing out of the mountains.

Jesse Frazier of Florence, near Cañon City, is credited with establishing the first large orchard, together with the first commercial nursery, in Colorado. He, too, received his first cuttings by wagon train from points east. Soon, he added his own root grafts, and in 1872 he harvested his first apples ("Ben Davis"). Within five years, his orchards yielded nearly 15,000 bushes annually, making him the largest apple producer of his day in Colorado. Frazier likely provided the starter stock to two notable Arkansas River valley orchardists, Captain B. F. Rockafellow of Cañon City and railroader J. H. Crowley of Rocky Ford.[20]

Recalling that the earliest attempts at fruit culture in northern Colorado were limited and only marginally successful, it is remarkable that by the mid-1870s commercial orchards were flourishing in the valleys of Clear Creek and the St. Vrain, Big Thompson, and Cache la Poudre rivers. Part of that prosperity undoubtedly had to do with the availability of a ready market because of an increase in population; a greater part resulted from the persistence and diligence of the orchardists themselves. Joseph Wolff of Boulder County, for example, had lost all his 300-plus fruit trees to cold, dry winds during the harsh winter and spring of 1872–1873. Yet, he was not discouraged. "Fruit culture in Colorado," he wrote, "is a system of experimenting, and must for many years be largely in that condition, until experience shall determine what varieties to plant, the soil required, the proper tillage, the effect of irrigation, mulching, fertilizers, and other equally as important matters."[21] Such sentiments were frequently repeated throughout the coming decades in both farm newspapers and the proceedings of the new Colorado State Horticultural Society.

Organized under the auspices of the *Colorado Farmer* newspaper at Denver on September 30, 1880, the society began with eighteen members from Front Range counties. By the time of its first annual meeting in 1882, membership had grown to thirty-nine men and seven women; it remained roughly at that level until the society's dissolution in 1922. Later, in 1944, the society would be reorganized as the Western Colorado Horticultural Society, signifying the shift of orchard production from the urban Front Range to the still rural Western Slope.

From its beginning, though, the Colorado State Horticultural Society's stated reason for being was broader than just fruit culture: it was the promotion of "horticulture, pomology, arboriculture and floriculture." The names of the initial standing committees reflected that horticultural breadth: Meteorology, Entomology, Ornithology, Geology, Forestry, Pomology, Vegetable Culture, Floriculture, and Ornamental Gardening.[22] Nonetheless, fruit culture retained central attention. The newer technique, for example, of grafting an apple twig or scion onto a root, rather than using the more common technique of inserting the scion into a slit made on the main stem of a tree, proved a more effective way to acclimatize non-native stock to Colorado: "It consists simply in grafting on to the root of a year-old seedling a scion or sprig from a standard apple of the same year's growth. The resulting tree is always true to the variety from which the scion was taken, and comes into bearing at a comparatively early period."[23]

In 1883, three years after its establishment, the Colorado State Horticultural Society obtained recognition by the legislature as the State Bureau of Horticulture, "to encourage and assist in the organization of district and county societies, and give them representation in the State Bureau, and in every proper way encourage and further the fruit and tree growing interests of the State." Given an annual appropriation of $1,000, the bureau began to publish an annual report that included horticultural statistics and essays with recommendations of practical utility to Colorado horticulturists.[24]

Arguably more active than the state society, the Northern Colorado [District] Horticultural Society covered Weld (Greeley), Larimer (Fort Collins), and Boulder counties. Its membership, nearly fifty, included representative of the first generation of Colorado fruit growers, among them Albert E. Gipson

(1848–1937) of Greeley and James S. McClelland (1837–1902) of Fort Collins, the first northern Colorado nurserymen and earliest promoters of an agricultural experiment station at Colorado Agricultural College in Fort Collins.

Gipson and McClelland were officers of the Colorado State Horticultural Society; and McClelland also served on the Colorado State Board of Agriculture, the governing board of Colorado Agricultural College. Such overlapping membership was common to the state groups in Nebraska and Kansas, too. One gets the impression that everyone shared a certain earnestness, devotion, and cliquishness, which, in retrospect, we might view as old-fashioned, ineffectual, and stuffy. To be sure, the society members were among the most educated and progressive horticulturists, both rural and urban, and they sincerely sought practical answers to the most pressing questions of the time.

Because these societies pre-dated the federally supported agricultural colleges and experiment stations, their members had yet to benefit from regional research by professors in the natural sciences. Thus, the papers read at meetings, and the discussions that ensued, generally reported on members' own field trials and errors, combined with their classical opinions on the aesthetic and moral aspects of horticulture. The agenda of one three-day annual meeting of the Northern Colorado Horticultural Society, for example, included this mix of topics: Success in the Apple Orchard, Growing of Small Fruits, Plum Culture, Report on Pomology, Strawberries, Horticultural Irrigation, State Legislation, Benefits of Horticultural Meetings, Canning and Preserving Fruits, Aesthetics in Horticulture, Cultivation of House Plants, Vegetable Culture, Home Surroundings and Their Influence, New Fruits, Landscape Gardening, Relation of Poultry to Horticulture, Cross-Fertilization, The Cottonwood Tree, Progressive Horticulture, Celery and Cultivation of Asparagus, and Best Trees for Ornament and Profit.[25]

The State Horticultural Society meeting in convention pursued the same pattern of papers followed by discussions. The society also sponsored a "Question Box," so members could submit inquiries in advance on other topics and hear answers at special sessions, such as one in 1882 from which this excerpt is taken:

> What soil is best adapted to the cultivation of celery? Rich, sandy soil.
>
> Is cultivation of the cranberry possible in Colorado? No.
>
> What vegetable grown for the Denver market yields the most profit per acre? It depends. One member thought cucumbers, another thought parsnips, and still another cabbage.
>
> What is the profit per acre of raising strawberries? $200–$400.

> Is the pear tree sufficiently hardy in our climate to warrant its extensive planting? Yes, if careful.[26]

While tree planting as a matter of state policy existed in Kansas and Nebraska legislation before Colorado entered the Union in 1876, Colorado remains the only state on the High Plains to mention tree planting in its constitution. On preservation of forests, Article XVIII, Section 6 calls for the state legislature to enact laws to "prevent the destruction of, and to keep in good preservation, the forests upon the lands of the state, or upon lands of the public domain." On encouraging tree planting, Article XVIII, Section 7 enables the legislature to "provide that the increase in the value of private lands caused by the planting of hedges, orchards and forests thereon, shall not, for a limited time to be fixed by law, be taken into account in assessing such lands for taxation."

In 1886 the Colorado Legislature joined Nebraska and Kansas in providing rewards for tree planting: two dollars per 100 trees planted along ditches, fences, and public roadways, if the trees were set no more than one rod (16.5 feet) apart and were maintained in good growing condition. Toward that end, *Field and Farm* recommended the "long-leaved" cottonwood (probably *Populus x acuminata* Rydb., a hybrid between *P. angustifolia* James found on the westernmost High Plains and *P. deltoides* Bartr. ex Marsh., the commonest poplar on the Great Plains) because it grew quickly and as a hybrid bore no cotton.[27] While it is always difficult to gauge the actual impact of state financial incentives on local economic activities, unquestionably since the 1870s, tree planting on the High Plains was "in the air." Nowhere was that more apparent than in the Union Colony of Colorado.

Notes

1. Everett N. Dick, *Conquering the Great American Desert: Nebraska* (Lincoln: Nebraska State Historical Society, 1975), 120.

2. Ibid., 118; John H. Hatton, "A Review of Early Tree-Planting Activities in the Plains Region," in *Possibilities of Shelterbelt Planting in the Plains Region*, edited by staff at Lake States Forest Experiment Station (Washington, D.C.: Government Printing Office, 1935), 54; Dorothy W. Creigh, *Nebraska, a Bicentennial History* (New York: W. W. Norton, 1977), 84; Kathryne L. Lichty, "A History of the Settlement of the Nebraska Sandhills" (master's thesis, University of Wyoming, 1960), 54.

3. Nebraska State Board of Agriculture Annual Report (1870): 3–7 (quote on p. 7); William H. Alderman, *Development of Horticulture on the Northern Great Plains* (St. Paul: Great Plains Region, American Society for Horticultural Science, 1962), 93.

4. James C. Olson, *J. Sterling Morton, Founder of Arbor Day* (1942; repr. Lincoln: Nebraska State Historical Society, 1972), 197–199.

5. Charles S. Harrison, "Essay on Tree Culture," Nebraska State Board of Horticulture Annual Report (1872): 51, 57

6. Nebraska State Board of Horticulture Annual Report (1877): 86.

7. Quoted in Olson, *J. Sterling Morton*, 162.

8. Ibid., 163.

9. J. Sterling Morton, "The Origin of Arbor Day and Its Results," Nebraska State Horticultural Society Annual Report (1902): 63.

10. Ibid., 64; Olson, *J. Sterling Morton*, 163–166.

11. Walter Prescott Webb, *The Great Plains* (1931; repr. Lincoln: University of Nebraska Press, 1981), 412.

12. Dick, *Conquering the Great American Desert*, 125.

13. Ibid., 119; Edwin A. Curley, *Nebraska 1875: Its Advantages, Resources, and Drawbacks* (Lincoln: University of Nebraska Press, 2006), 352–353.

14. U. P. Hedrick, *A History of Horticulture in America to 1860* (New York: Oxford University Press, 1950), 332.

15. E. R. Ware and Lloyd Smith, "Woodlands of Kansas," *Kansas State Agricultural Experiment Station Bulletin* 285 (July 1939): 7.

16. *Kansas Farmer* 4, no. 11 (November 1867): 176; George A. Filinger, "The Kansas State Horticultural Society, 100 Years of Progress, 1867–1967" (mimeograph) (Manhattan: Kansas Horticultural Society, 1968), 4.

17. Richard S. Elliott, "Farming in the Great American Desert," *Kansas Farmer* 7, no. 11 (November 15, 1870): 173; *Kansas Farmer* 9, no. 15 (August 1, 1872): 233–234.

18. John H. Tice, *Over the Plains, on the Mountains; or, Kansas, Colorado, and the Rocky Mountains; Agriculturally, Mineralogically and Aesthetically Described* (St. Louis: "Industrial Age" Printing, 1872), 49, 250; David M. Emmons, "Richard Smith Elliott, Kansas Promoter," *Kansas Historical Quarterly* 36, no. 4 (Winter 1970): 390–401.

19. Martin Allen, "Horticulture on the Plains," *Kansas Horticultural Report* 8 (1878): 189.

20. Alvin T. Steinel, *History of Agriculture in Colorado* (Fort Collins: State Agricultural College, 1926), 501–502, 511.

21. Quoted in William E. Pabor, *Colorado as an Agricultural State* (New York: Orange Judd, 1883), 164–165.

22. Colorado Horticultural Report 1 (1884): 17–24.

23. W. F. Watrous, in Colorado State Horticultural Society Annual Report (1884): 218.

24. Ibid. (1885): 15.

25. Ibid. (1886): 385–584.

26. Ibid. (1884): 30.

27. *Field and Farm* 1, no. 19 (May 8, 1886): 7.

3

Horticulture for Home and Community

Among the very first purchases for the Union Colony of Colorado, ordered by its founder, Nathan C. Meeker (1817–1879), in April 1870, was a railroad car full of shade and fruit trees from the Bloomington Nurseries. Since irrigation water was not yet available at the Greeley town site, the trees were temporarily heeled into trenches, with roots well covered, close to the Cache la Poudre River. It was not until June 1871, when the first irrigation canal brought water into town, that the trees could be planted permanently. Because the early growing season had passed and there was not enough time for solid rooting before the advent of cold weather, all the fruit trees and most of the shade trees were dead by the following spring.

Meeker readily acknowledged, four years later in his *Greeley Tribune*, that when he and the other colonists first located in Colorado Territory, "we had no kind of idea of the difficulties attending the culture of many kinds of vegetables. The great variety of forest trees which grow in the states without any trouble, many of them as spontaneously as weeds, can here scarcely be made to live when brought hither with the greatest care and cultivated with the utmost attention." And yet by 1890, Captain David Boyd, among the first Union colonists, described Greeley, population 2,500, as a community of homes surrounded by fine evergreens, deciduous shade trees, and lawns.[1] Nevertheless, this was not precisely the community Meeker had planned; nor had horticulture developed quite as he had envisioned.

Meeker's idealism, even if brass-bound, presaged some present-day notions for using horticulture as a means of creating a sense of place. The steps in his career that took Meeker to Greeley illustrate the intertwining of the practice of cultivating the soil strictly speaking and the more generally civilizing aspect of horticulture in the settlement of the High Plains. It is not too much to say that the High Plains owe much to the example of Meeker and his fellow colonists, in the sense that they helped provide "an impulse" to the development of horticulture as an expression of civic-mindedness.[2]

In 1834, at age seventeen, Meeker left the family farm at Euclid, Ohio, to seek his literary fortune in New York City. After eight years without achieving success, he took a teaching job in Long Hill, New Jersey. Continued meager earnings may have been the immediate cause of his attraction to the ideas of the French utopian socialist Charles Fourier, which Meeker found described in a periodic column of Horace Greeley's *New York Tribune*. After less than a year of teaching, Meeker returned to Ohio where he began to write and lecture on Fourier.

Like many utopians before and since, Fourier was quixotic: idealistic and a bit odd. He had adopted the notion that social harmony could best be achieved through the establishment of small cooperative communities. Each individual entering such a community, known as a phalanx or phalanstery, would obtain landed property, thereby securing a stake in the community and becoming an equal with other property holders. As individuals cooperated and competition disappeared, harmony and happiness would result, all, of course, based on the assumption of the ultimate goodness of human nature.

While the first Fourier community had been established near Paris in 1832, the movement enjoyed its greatest popularity in the United States during the 1840s, when sixteen communities were founded—most notably Brook Farm near Roxbury, Massachusetts. Less well-known was Trumbull Phalanx at Braceville, near Warren, Ohio, of which Meeker was a founding member and the corresponding secretary. Trumbull started with a few primitive log buildings on a 275-acre plot and not as the community "palace" Fourier had envisioned or Meeker had expected. The phalanx did, however, come close to self-sufficiency in growing its own food—a menu based on whole wheat flour and abstinence from alcoholic drinks. Meeker and his young family resided at Trumbull from 1844 until 1847 when a malarial fever, the ague, caused the phalanx's gradual debilitation and dissolution.³

Again close to destitute, Meeker joined his family's general merchandise business in Euclid for two years before accepting an invitation to open a general store in Hiram, Ohio, where the Campbellites, an ecumenical offshoot of the Disciples of Christ, were building a new college; he was probably baptized into that denomination. At Hiram, Meeker continued to lecture on Fourier, and he developed his own ideas about community, agriculture, and religion. Meeker's son, Ralph, recalled a Mormon neighbor at Hiram who spoke in glowing terms about the beauties and resources of the Rocky Mountain region. Evidently, this stirred in Meeker père an interest in moving west to "found a community of a few families, removed from the noise and frivolities of society."⁴ As a follower of Fourier, Meeker believed society and not human nature was the source of all corruption and vice.

Around 1857, again near bankruptcy, Meeker and his family moved to southern Illinois, attracted by the salesmanship of Stephen A. Douglas and the Illinois Central Railroad, to establish a small farm and general store. While growing vines—strawberries in particular—for the Chicago market, Meeker continued his newspaper writing, both agricultural and abolitionist. That attracted Horace Greeley, who hired him as war correspondent assigned to General Ulysses Grant and, after the Civil War, as agriculture editor. In the latter capacity, Meeker wrote a distinguished series of articles on Oneida, the utopian community in western New York. In the fall of 1869, Greeley sent Meeker to the Rocky Mountain West, with Utah as the final destination. Stories about Mormon rituals, especially communal irrigation, had intrigued Greeley. He wanted Meeker to prepare a series of articles on Mormon communities and their agriculture.⁵

Meeker traveled by rail as far as western Kansas to the end of the rail line, then by horse up the Arkansas River valley via Fort Lyon and Bent's Fort, then north to Denver, which, because of its recent prosperity, Meeker compared favorably to the finest towns of Illinois and Ohio. While an early fall snowstorm prevented Meeker from reaching Utah, he did explore the Front Range as far north as Cheyenne before returning to New York.

On December 14, 1869, Meeker published the first notice, endorsed by Greeley, setting forth both the plan for "A Western Colony" (the exact location was left unmentioned to avoid attracting land speculators) and the qualifications for colony membership: "The persons with whom I would be willing to associate must be temperance men, and ambitious to establish good society." They also needed to be men of some means who were willing to pay an immediate $5 initiation fee, to cover a $150 membership fee to help defray the cost of selecting and purchasing land for the colony, and to possess sufficient savings to support their families during the colony's early period. Each member was entitled to land for farming outside town and the right to buy a town lot for $25 to $50. Meeker specifically called for horticulturists—farmers, nurserymen, and florists—and, generally, for practitioners of those professions and occupations that made for "an intelligent, educated, and thrifty community."[6]

Within less than a month, Meeker had received over 1,000 letters of interest from throughout the nation. At an organizational meeting on December 23, 1869, in New York City, he was elected president of the Union Colony; retired general Robert A. Cameron from upstate New York, vice president; and Greeley, treasurer. By December 27, an executive committee of eight members, including the elected officers, had drawn up a constitution and bylaws, primarily for the purchase and distribution of land. Once the locating committee found a suitable site, the treasurer meant to purchase and hold that land in trust for the colony; the executive committee was authorized to deed parcels of land to members upon receiving verification that they had made improvements on their respective parcels.

Meeker's town plan for Greeley imitated that of the New England village. The town site, approximately 640 acres, consisted of business and residential lots with some land reserved for a commons, schools, churches, and other public institutions. The lots were laid out so they could "be sold, one of each to each member of the Colony, at a fixed valuation, and the proceeds devoted to improvements for the common welfare." In addition, tracts surrounding the town site were "to be divided into lots of 5, 10, 20, 40, and 80 acres, according to their distance from the town center, and deeded one to each member."[7] Since some of these surrounding tracts were more desirable than others, Meeker suggested selling them by auction, with proceeds again going for community improvements. Meeker conceived of the Union Colony as a cooperative venture only with regard to the initial purchase and distribution of land; all other transactions were free of community control.

Meeker had found the New England village plan attractive because it provided all residents with convenient access to businesses, schools, and public places and thus created a virtually immediate sense of community. Moreover, the New England plan encouraged horticulture because "in planting, in fruit growing and improving homes generally, the skill and experience of a few will be common to all, and much greater progress can be made than where each lives isolated."[8]

Nathan Meeker's Greeley, 1870. Courtesy, City of Greeley Museums, Permanent Collection.

Meeker considered a comfortable and attractive home not simply a pleasant amenity but a necessary prerequisite for preserving the family. His view of the connection among family, community, and horticulture came out in the summary of aims and values of the *Greeley Tribune*, which he launched on November 16, 1870: "to enforce the doctrine that the foundation of all prosperity, whether of nations or individuals, is based on the family relation as maintained in civilized countries, and that the highest ambition of the family should be to have a comfortable and, if possible an elegant home surrounded by orchards, and ornamental ground, on lands of its own." Meeker hoped the success of the Union Colony would serve as a model (his word) for establishing family-oriented communities in the remainder of the unsettled West.[9]

Meeker had provided the general plan for the colony; he left its details to the Union Colony of Colorado, Inc. Its board consisted of the three-member locating committee (including Meeker), plus the corporation's Denver attorney and William Byers of the *Rocky Mountain News*. Byers also worked as an agent of the railroad from which much of the colony's land had been purchased. His exuberance about agricultural development in Colorado, contrary to Meeker's soft-spoken approach, may help explain why the Union

Colony of Colorado began to exaggerate the economic opportunities available at Greeley. Indeed, when the first colonists arrived in April–May 1870 and did not find what they had expected, those without confidence in Meeker personally or in his community-building ideas more broadly moved elsewhere.[10]

The crux of the matter was the lack of water. While Meeker recognized from the outset that water imported through canals and ditches was essential to the colony's success, the constitution written in New York made no mention of the related issue of water rights. Directors of Union Colony of Colorado, however, did establish the principle tying water rights to land ownership and required that canals and ditches providing water to the colony be built and managed by the colony administration, not by individual water users.[11] General Cameron estimated that $20,000 would cover the cost of four ditches to irrigate 120,000 acres. By the time the system was completed, although reduced to two ditches and covering less acreage, actual costs came to $412,000. Meanwhile, the value of water rights increased dramatically. Those rights tended to be oversold by the colony, leading to prolonged and costly litigation.

Because of unanticipated high costs, the common fund—consisting of proceeds from land sales, which Meeker had envisioned as paying for community amenities—had to be diverted to support the construction of canals and ditches.[12] In addition, colonists voluntarily raised monies through special subscription, and many also volunteered their labor for the construction of the canals. The first canal completed, known as Greeley #3, was a relatively short-distance, bottomland diversion from the south side of the Cache la Poudre River onto town lots and gardens. On the north side, Greeley #2, which measured thirty-two feet wide at its bottom, took out water several miles above town and ran for thirty-six miles. So far as is known, Greeley #2 became the first formal cooperative irrigation venture and the first primarily agricultural canal on the High Plains to provide for irrigated cultivation beyond stream valleys to benchlands.[13]

Despite the canals, Meeker's dream of a vital community surrounded by small-scale farms producing a variety of vegetables, fruits, and nursery stock would not come true. Instead, Greeley developed as a center of large-scale, single-crop agriculture. Irrigated benchlands eventually became consolidated into larger and larger fields for the cultivation of potatoes, rotated with the newly popular soil restorative and fodder crop, alfalfa (*Medicago sativa* L.). Also called lucerne, seeds of this plant, native to southwest Asia, were imported in great quantities; it is now naturalized throughout North America. By the end of the nineteenth century, approximately 30,000 acres around Greeley

Lateral from Ditch #3 entering Greeley, 1870. Courtesy, City of Greeley Museums, Permanent Collection.

were alternatively planted in alfalfa and potatoes, primarily the late-season red potato known as the "Greeley spud." Indeed, the planting of alfalfa led to the development of feeder operations that fatten up livestock for the markets and still emit a noticeable scent into the atmosphere around Greeley.[14]

At the very beginning, the presence of cattle around Greeley had caused the colony to hire herders to keep animals off cultivated areas. Rules were enacted but ignored by the early colonists who ran cattle, leading Meeker to editorialize that the issue was "not so much a question in regard to fences as in regard to order and decency, for our town and colony will be disgraced by cattle at large running through our streets; shade trees will be impossible, for even fences themselves will be comparatively useless since there are enough breachy [sic] cattle to demolish common fences."[15]

During the spring of 1871, the colony administration did use some common fund monies to construct nearly fifty miles of fences to enclose its lands. Gaps remained, however, because the Colorado Territorial Legislature, pressured by the cattle interests, had prohibited closed gates across public roads

Lincoln Park, Greeley, ca. 1875. Courtesy, City of Greeley Museums, Permanent Collection.

each year during the months of May through September. In the stockgrower's view, according to Captain Boyd, the High Plains lent themselves only to grazing; thus, those wishing to do agriculture or horticulture were fools: "Our fence was ridiculed, and we were accused of being proud and wanting to keep ourselves to ourselves, as a peculiar and very holy people. We were Greeley 'Saints' who had fenced ourselves in from the 'heathen around about.'"[16]

Actually, quite the opposite had occurred. During the first year, colony farmers did not keep to themselves. They established the first Farmers' Club in Colorado, consisting of about sixty members who met every Wednesday evening to discuss farming techniques. The Greeley Farmers' Club sponsored the first Farmers' Institute on the High Plains in December 1870 and, within three years, would spawn at least nine similar clubs along the Front Range. Already well established in the older states, the Farmers' Institute provided instructional sessions on agriculture and horticulture and served as the precursor to the extension activities of the agricultural experiment stations.

The topic of the first Farmers' Institute at Greeley was tree planting. Nathan Meeker spoke in favor of planting fruit trees; his fellow officer in the Farmers' Club, J. Max Clark, discoursed on hardy fruits such as apples

Greeley Nursery products, 1890s. Courtesy, City of Greeley Museums, Permanent Collection.

and sour cherries.[17] Naturally, the most frequent topic of discussion at weekly meetings of the Greeley Farmers' Club was irrigation, the quantity of water needed to irrigate specific areas and how best to apply it. "From the farmers themselves," Captain Boyd reported, "we learned that much more water was needed than the theorists writing for newspapers had been endeavoring to make us believe," which "led us to investigate the whole subject in concert, and to work together for the attainment of certain ends."[18]

As president of the Greeley Farmers' Club and later as a member of its successor, the Grange, Boyd had a crucial hand in the many discussions that ultimately led to the creation of the Colorado system of water-rights enforcement. In 1879 he presided over the select committee that drafted the Colorado legislation. Suffice it to note that the Greeley colonists, living and farming on the arid plains many miles beyond the foothills, had recognized very early that their claims on Cache la Poudre River water would be uncertain until all claims to that water were defined, made a matter of written record, and then enforced by the appropriate authorities.

While legislation on water rights came as a result of conflicts on the Cache la Poudre River during the 1870s, Meeker had observed the use of water supplied by means of ditches in Denver on his first trip west in 1869. A city with a population already near 4,000, Denver was a place, according to Meeker, where "rain seldom falls, and cultivation would be without reward were it not for the introduction of water from the [South] Platte, which, being taken out of this stream, twenty-four miles above the city, runs along the gutters of every street, and into gardens."[19] Meeker was apparently unaware that four years earlier, in 1865, the first ditch to reach the city site was dug from Cherry Creek, approximately where the present-day Speer Boulevard intersects Broadway.

Indeed, William Byers, also a director of Union Colony of Colorado, Inc., claimed to be among the first private citizens to use the Cherry Creek water. He had transplanted cottonwood (probably *Populus deltoides* Marsh.) from locations unknown to his yard on Arapahoe near 15th Street and irrigated the saplings by hand, carrying water by bucket from the ditch. To those who found the "cotton" (cottony hairs discharged from seed caps) annoying, Byers countered that "cotton" flew only during a very short period in summer and that the cottonwood, being native, was the most readily available and fastest-growing shade tree for Denver and thus needed to be planted widely.

In later years, Byers reflected that the boxelder (*Acer negundo* L.) followed the cottonwood in both his personal estimation and its overall popularity, although it, too, came under criticism because it tended to be worm-infested. Byers also liked the black locust (*Robinia pseudoacacia* L.), although a non-native; the Russian-olive (*Elaeagnus angustifolia* L.) for its hardiness and quick growth; and the "wild black cherry," by which he probably meant the native chokecherry (*Prunus virginiana* L. var. *melanocarpa* (A. Nels.) Sarg.) rather than the non-native wild black cherry (*Prunus serotina* J.F. Ehrh.), for attracting birds.[20]

Like Byers, William E. Pabor (1834–1911) had made the connection between horticultural and community development, both in the Union Colony and in Denver. A native of New York City and among the original members of the Union Colony, Pabor, too, was a publicist (he prepared the Union Colony's first annual report), realtor, and amateur horticulturist. To readers of the *American Garden* (November 1882), Pabor described his "Garden Culture by Irrigation," a four-acre plot at Shadyside, his residence two miles from downtown Denver. He had begun with the tap of a nearby irrigation canal: he fitted the tap with a wooden headgate, made to lift so as to allow the necessary amount of water to flow under it and directly onto a channel or flume lined with wood on the bottom and sides, and then into a main ditch. From that ditch, Pabor ran several smaller ditches, or laterals, to his various horticultural beds. He could open or close the laterals as needed, simply by moving two or three shovelfuls. Each bed consisted of irrigation furrows and was surrounded by an earthen mound to further control the flow of water. Among the various sections, Pabor cultivated a "kitchen garden," with twenty-plus varieties of vegetables; asparagus and rhubarb beds; beds for small fruits including currants, raspberries, and strawberries; and an ornamental flower garden. He also maintained "a sort of experimental garden, where new and choice varieties of seeds are tested."[21]

Although the extent of his experimentation is unknown, it is possible that Pabor obtained at least some seeds from the Colorado Agricultural College. The State Board of Agriculture had permitted the college to distribute, free of charge to citizens, seeds not used in experimentation, provided that recipients reported their planting results back to the college horticulturist.[22]

To be sure, the gardens at Shadyside were exceptional in both size and refinement. By the 1880s, the Colorado State Horticultural Society had recorded numerous examples of more ordinary gardens, both in communities and on farms—testimonials as to what flourished and what did not under unique local conditions. At its 1882 annual meeting, Mrs. A. L. Washburn read a paper on the "Ideal Garden," addressing the questions of how and when to irrigate and which were the best fertilizers to use in vegetable and flower gardening. And she posed additional questions: Why do certain vegetables, such as melons and turnips composed chiefly of water, require less water than many others? Is it best to run water close around tree trunks or at a little distance? Has anyone tried lime or gypsum as fertilizer?[23]

Commenting on the moral or cultural value of horticulture at a later meeting, Mrs. M. D. Cole of Berthoud noted that there was much more to gardening than deciding which varieties to plant and how they should be cultivated. In "Woman's Work in Horticulture," she appealed to farm and ranch women to spend less time "scrubbing and ironing" and more time working outdoors in the garden: "Our health and our nerves demand it, and for our digging and working the soil, mother earth will well reward us, not only in dollars and cents, but she gives a bonus besides, of renewed vigor and strength." She acknowledged that farm and ranch women still endured many privations, although thanks to modern appliances, they were slowly gathering the comforts of life.[24]

While acknowledging that, during springtime, men and boys on stock farms needed to occupy themselves with sowing forage crops, Mrs. Cole affirmed that they still could be relied upon to plow and harrow garden plots. Thereafter, she encouraged women to take up a new routine by going into their gardens immediately after breakfast—leaving housework undone for an hour or more—to breathe in the cool of the morning, to clear their minds of drudgeries, and to work intently and thoughtfully among their plants. She admonished them to take their children: "Those not old enough to help will enjoy making mud pies. Indeed, I pity any child who is denied the privilege of making mud pies. And, above all, let us inspire our boys and girls with a love of nature and of agricultural pursuits, if you would keep them on the farm."[25] Virtually since the beginning of settlement, retaining young people has proven among the most difficult challenges on the High Plains.

Underlying all challenges, however, is the basic fact of limited water. It is good to keep reminding ourselves that on the High Plains, scarcity of moisture is far more common than plenty of moisture and that in the period before storage reservoirs, trans-basin diversion, and deep wells, prolonged drought was truly devastating. When such a drought returned to the region in the late 1880s, those living beyond the communities of the Front Range and who could maintain even an acre or two of irrigated garden were able to remain on their homesteads. Indeed, if there is a single symbol of human existence on the more isolated sections of the High Plains, it is the windmill.

Stockgrowers introduced windmills onto the High Plains to operate the pumps that ensured continual and reliable supplies of water for their stock. The railroads used windmills to fill their water towers, and homesteaders began using them in the 1880s. In a highly original paper, Erwin H. Barbour (1856–1947), professor of geology at the University of Nebraska and acting state geologist, documented the value of windmills to farmers and ranchers along the Loup, Republican, and Platte River valleys in western Nebraska. He was intrigued by their homemade construction—consisting of parts from old machines, scrap iron, and wood—and was impressed by their low maintenance costs. And this, combined with his belief in an unlimited supply of groundwater, convinced him that, when the rains and snows failed, the homesteaders could always rely on the wind.

Beyond guaranteeing the necessities for existence, the windmill could ensure those amenities that made the difference between simple survival and a more comfortable life. "The sight of a sod house with flower beds and a lawn sprinkler," Barbour wrote in 1899, "is unexpected and almost incongruous. Very different from the prevailing idea of frontier life are hot and cold water faucets. With the picture of cowboy life as the half-informed press paints it one would never expect to find a ranch house with marble basins and porcelain tubs. Such things exist, and are due wholly to the agency of the wind utilized by the windmill."[26]

To this day, as one drives the highways and byways of the High Plains, the windmill remains the most visible sign of human presence. And even on the countless homesites abandoned for whatever reason, the windmill remains, at

Operating windmill near Lingle, Wyoming. Courtesy, Ronald K. Hansen.

least in its severed parts, to remind the passerby that people once lived here. The windmill has left its mark on local mythology, too, serving as symbol for the ingenuity required to thrive on the arid plains and reminding us that the wind remains an ever-present source of renewable energy.[27]

While the windmill still serves as an efficient provider of water to livestock on more isolated sections of land, whether public or private, its presence did not preclude cultivation of the land. Yet during the 1860s and 1870s, when large nomadic cattle operators had taken control of watered valleys, especially in western Nebraska and eastern Wyoming, they effectively prevented cultivation. These nomadic operators viewed the Homestead Act and related legislation as part of a deliberate federal effort to deprive them of access to those public lands they believed were best suited for grazing.

Their leading advocate, the Wyoming Stock Growers Association (established 1879), came to wield extraordinarily effective political clout through the Wyoming Territorial Legislature and, after 1890, through the state legislature and the Wyoming congressional delegation. It is quite remarkable, therefore, that Governor John W. Hoyt managed to get the 1881 Wyoming Territorial Legislature to adopt a resolution urging Congress to change laws so that watercourses would be "available for the common and equal benefit, so far as possible, of all persons who may choose to occupy and improve any part of the public domain." That same year Governor Hoyt reported to the secretary of the interior that "great numbers of ranchmen" had successfully grown crops of potatoes, oats, and even wheat—an early indication of that combination of animal husbandry, agriculture, and horticulture that became known as the stock farm.[28]

By no means were all stockgrowers opposed to agriculture or horticulture. In fact, the owner of one of the largest ranches in Wyoming turned out to be perhaps the most public enthusiast for horticulture. Joseph M. Carey (1845–1924), a native of Delaware and a law graduate of the University of Pennsylvania, was appointed U.S. district attorney for Wyoming Territory by President Grant. Shortly after reaching Cheyenne in 1872, Carey started to grow potatoes and other vegetables on his property north of town; not one year elapsed, not even when he represented Wyoming in Congress or served as governor, without him taking a personal interest in gardening.[29]

Territorial governor Hoyt had already noted that within a very short period of time, hundreds of new home gardens came under cultivation in Wyoming. In 1887 one of his successors, Thomas Moonlight, drew special attention to gardening successes on the Laramie Plains at elevations exceeding 7,200 feet. He reported on Mayor Trabing of Laramie who farmed sixty acres of irrigated (and fenced) land on the west side of the Laramie River. Trabing raised and sold potatoes, turnips, beets, onions, peas, and cabbages, as well as wheat, oats, barley, and alfalfa.[30]

While we have noted the existence of vegetable gardening by the middle of the nineteenth century at Fort Laramie, the first indication of ornamental planting in Wyoming Territory came well after the Civil War. Post records indicate that a hospital flower plot existed in 1869, cared for and watered every day by the patients. By 1878 the military had expanded the plot to include four square flower beds with a raised, round flower bed in their center. Apparently, the assistant post surgeon supervising the planting had some knowledge of plant classification. While some of his flower seeds had been purchased, several had been collected from plants growing locally: larkspur (probably *Delphinium nuttallianum* Pritz. ex Walp.), purslane (probably *Portulaca oleracea* L.), corn cockle (*Agrostemma githago* L.), pink (*Dianthus armeria* L.), alyssum (*Alyssum alyssoides* L.), wallflower (probably *Erysimum asperum* (Nutt.) DC.), mignon mirabilis (*Reseda lutea* L.), field bindweed (probably *Convolvulus arvensis* L.), and vervain (probably *Verbena stricta* Vent.).[31]

Fort Laramie sits on a rise a few hundred yards from the Laramie River. Thus, one can safely surmise that when the first trees were planted along the parade grounds in 1877, they came from the nearby bottomlands. The military also provided for the first trees to be planted in the new city of Cheyenne, seventy-five miles to the south. Located in a slight depression on the plains that lead from an elevation of 6,000 feet to the foothills twenty miles west, Cheyenne was treeless, except for cottonwoods and willows along nearby Crow Creek. In 1877, however, the supply officer at Fort D. A. Russell (now Fort Warren) procured "20,000 young trees and willow cuttings" from an unnamed Nebraska nursery as a gift to Cheyenne, which was less than two miles from the fort. Because of the transcontinental railroad, Cheyenne's population quickly grew to 5,000.[32]

With delivery of the Nebraska trees, Cheyenne residents inaugurated an organized campaign, which continues to the present day, to make Cheyenne a "City of Trees." Leading that effort was a grocer named James F. Jenkins, who years later recalled how his business partner and a local jeweler used to visit in the grocery store, "and we talked about tree planting in Cheyenne, as Nebraska was doing so much in that line of development."[33] Looking for opportunities to plant, the businessmen knew the Union Pacific had gifted four blocks of company land to the city for park purposes and that nothing had yet been done to improve that land. Subsequently, the jeweler volunteered to "go out with a subscription paper and obtain money to start a park, on Arbor Day, 1882."[34]

The subscription was successful, and trees—mostly cottonwoods—were secured for transplanting. Arbor Day 1882 was "a gala day," Jenkins reminisced: "Everybody got in line and paraded the streets. Mayor John Talbot dug up a nice tree and had it loaded on a Government wagon, drawn by eight black mules. . . . The Mayor had placed a flag on the top of the tree, and after a good tramp through the streets, it was planted in the Park." Residents were invited to purchase trees not only for the park but also for their own yards. In the park, trees were planted in rows twelve feet apart each way, and irrigation was provided by means of an earthen ditch from Sloan's Lake. Since the park began as little more than open prairie, subscription monies were also used to build a fence "to keep the big City herd of milk cows from destroying [the trees]."[35]

In view of citizen interest in tree planting and community beautification generally, the Cheyenne City Council created the post of park commissioner in 1883 and appointed Henry Altman, one of its members, to fill the position. Once Altman began working at the park, leveling the land and making other improvements, he aroused the ire of a city council member who opposed the expenditure of public monies for the park. As a result, Altman did what quiet but determined citizens have always done—simply used his own money to purchase more trees, flower seeds, and other items to complete the park.[36] By the 1890s, Cheyenne had indeed become a "City of Trees," with plantings throughout residential areas and along the main road to Fort Russell. The city's further development as a horticultural oasis depended in great part on the collecting, storage, and transportation of distant waters. This required new water laws and regulations, as well as substantial federal investments. The foundation for such changes had been sketched out by John Wesley Powell in 1878 but would not be implemented until after the turn of the century.

Notes

1. David Boyd, *A History: Greeley and the Union Colony* (Greeley: Greeley Tribune Press, 1890), 79–81.

2. William E. Smythe, "Real Utopias in the Arid West," *Atlantic Monthly* 79 (May 1897): 604.

3. Lowell Baumunk, "Nathan Cook Meeker, Colonist" (master's thesis, Colorado State College of Education, 1949), 16–25.

4. Boyd, *History of Greeley*, 15.

5. Baumunk, "Nathan Cook Meeker," 27–31.

6. Nathan Meeker, "A Western Colony," in James F. Willard, ed., *The Union Colony at Greeley, Colorado, 1869–1871* (Boulder: University of Colorado Historical Collections, 1918), 1–4.

7. Quoted in Baumunk, "Nathan Cook Meeker," 39–40; Willard, *Union Colony*, xix–xxv.

8. Quoted in Boyd, *History of Greeley*, 33.

9. Ibid., 32–40, 83.

10. Ibid., 43–49.

11. Colony certificate of organization in Willard, *Union Colony*, 15–16.

12. Donald J. Pisani, *To Reclaim a Divided West: Water, Law, and Public Policy, 1848–1902* (Albuquerque: University of New Mexico Press, 1992), 80.

13. Robert G. Dunbar, "History of Agriculture," in Leroy Hafen, ed., *Colorado and Its People* (New York: Lewis Historical Publishing, 1948), 2:122; Dunbar, *Forging New Rights in Western Waters* (Lincoln: University of Nebraska Press, 1983), 22.

14. Minutes of Union Colony, June 18, 1870, in Willard, *Union Colony*, 44; J. Max Clark, "Potato Culture Near Greeley, CO," USDA Annual Report (1904): 311, 316.

15. Nathan C. Meeker, *Greeley Tribune* Extra, n.d. [October 1870], quoted in Boyd, *History of Greeley*, 69.

16. Boyd, *History of Greeley*, 68–70.

17. Alvin T. Steinel, *History of Agriculture in Colorado* (Fort Collins: Colorado State Agricultural College, 1926), 67–71.

18. Boyd, *History of Greeley*, 267–268.

19. Ibid., 25.

20. William N. Byers, "Shade Trees," Colorado State Board of Horticulture Annual Report 10 (1898): 86–88.

21. William E. Pabor, *Colorado as an Agricultural State* (New York: Orange Judd, 1883), 210.

22. Frank J. Annis, "Concerning the Duties of the Secretary of the State Board of Agriculture, and the Distribution of College Seeds and Plants," *Colorado Agricultural Experiment Station Bulletin* 3 (December 1887): 1–4.

23. Colorado Horticultural Report 1 (1884): 31.

24. Ibid., 3 (1886): 224–229.

25. Ibid.

26. Erwin H. Barbour, "Wells and Windmills of Nebraska," *Water-Supply and Irrigation Papers of the United States Geological Survey* 29 (1899): 32.

27. Walter Prescott Webb, "The Story of Some Prairie Inventions," *Nebraska History* 34, no. 4 (December 1953): 232–233.

28. Annual Report of Wyoming Territorial Governor to Secretary of the Interior (1881): 62, 70.

29. Joseph M. Carey, "The Future of Horticulture in the State of Wyoming," *Wyoming State Board of Horticulture Special Bulletin* 1 (1907):21.

30. Annual Report of Wyoming Territorial Governor to Secretary of the Interior (1887): 4.

31. "Medical History" (spring 1878), unpublished compilation, Fort Laramie National Historic Site Library, Fort Laramie, Wyoming.

32. "Medical History" (May 1881) unpublished compilation, Fort Laramie National Historic Site Library, Fort Laramie, Wyoming; A. C. Hildreth, "Horticulture on the Wyoming Plains," in W. H. Alderman, ed., *Development of Horticulture on the Northern Great Plains* (St. Paul: Great Plains Region, American Society for Horticultural Science, 1962), 121.

33. James F. Jenkins, "My Life Story" [1925], copied by his daughter, Agnes Metcalf, from the original, Wyoming State Archives, Cheyenne, Jenkins Collection, MSS 7, 35.

34. Ibid.

35. Ibid., 35–36.

36. Ibid.

4

Toward "A New Phase of Civilization"

John Wesley Powell (1824–1902) is best known for his exploration of the canyons of the Colorado River. For the settlement of the West, however, his fame rests on his *Report on the Lands of the Arid Region of the United States* (1878). Although the book includes lands beyond our geographic area, his analysis and recommendations provide the frame of reference for the use and conservation of water, without which there could be no horticulture of any kind on the High Plains. Powell proposed new water laws and regulations completely different from those of the humid East, and he believed—incorrectly, it would turn out—that their implementation would guarantee that the arid West would become, and remain, a place of small farms and small towns. An admirer of the Union Colony of Colorado, and much like Nathan Meeker, Powell was both an idealist and a tragic figure when it came to the development of horticulture and community on the High Plains.

Powell published his *Report* at the very time thousands of settlers were attracted to the High Plains by the false claim that the kind of agriculture and horticulture practiced in the humid East could easily be extended to the arid West. In other words, the High Plains could be cultivated without irrigation, by rainfall alone. At the core of this false claim was a proposition, apparently first suggested in the early 1840s by Josiah Gregg—a trader on the Santa Fe Trail—and championed in the late 1860s by, among others, railroad agent Richard Elliott in western Kansas. "The high plains," Gregg wrote, "seem too dry and lifeless to produce timber; yet might not the vicissitudes of nature operate a change likewise upon the seasons? Why may we not suppose that the genial influences of civilization—that extensive cultivation of the earth—might contribute to the multiplication of showers, as it certainly does of fountains? Or that the shady groves, as they advance upon the prairies, may have some effect upon the seasons?" Gregg based his seemingly sentimental view on the opinions of older settlers who had reported to him that droughts were "becoming less oppressive in the West" and that "the rains have much increased of latter years, a phenomenon which the vulgar superstitiously attribute to the arrival of the Missouri traders."[1]

Starting in the 1870s, two outspoken "boomers," Samuel Aughey and Charles Dana Wilber of Nebraska, elaborated on Gregg's thesis, citing indisputable "facts" to conclude that, as human settlement moved westward, rainfall had increased. Aughey held the titles of professor of natural sciences and state geologist at the University of Nebraska; Wilber was employed by the university's Department of Geology and Mineralogy and was a member of the Nebraska Academy of Sciences. Both worked as advisers to the railroads, which used their promotional writings well into the twentieth century.

In 1880, Aughey brought up several "facts of nature" to demonstrate the trend toward increased rainfall: reappearance of spring-fed creeks that had been dry since 1864, taller stands of grasses in the 1870s than observed by Lewis and Clark in 1804, and precipitation reports from various Nebraska localities (including western Nebraska) that documented greater rainfall during the period 1869–1879 than in 1859–1869. As to what caused those changes, Aughey dismissed what today might be considered the most likely reason, namely, the cyclical nature of climate. "Independent of any human agency," he wrote, "there are no cosmical [sic] causes definitely known that would cause an increase in rainfall over an isolated region [Nebraska] of the earth." He rejected tree planting as a "main cause" but accepted it as a "helping cause." He had advocated tree planting in a message to the Nebraska Legislature in 1874, the year following passage of the Timber Culture Act. In addition, he rejected the notion, still popular in some quarters, that the steel used in railroad and telegraph lines had somehow increased the amount of electricity in the atmosphere, which, in turn, had caused more rainfall. Older states with more railroad and telegraph lines that had existed for a longer period of time than was the case in Nebraska had experienced no increase in rainfall.[2]

The main cause, Aughey concluded, was "the great increase in the absorptive power of the soil, wrought by cultivation, that has caused, and continues to cause an increasing rainfall in the State."[3] According to Wilber, since 1867, Aughey had conducted many "experiments" to substantiate that conclusion. For ages, Wilber continued, the soils of Nebraska had been "pelted by the elements and trodden by millions of buffalo and other wild animals, until the naturally rich soil became as compact as a floor." With the advent of settlers, the soil became broken by the plow, allowing rainfall to be absorbed. As the soil slowly returned this moisture to the atmosphere and more soil was cultivated, more moisture evaporated, resulting in increased rainfall.[4]

It was Wilber who coined the famous phrase "rain follows the plow." He did this as a way of summarizing his studies and of explaining that as settlements moved westward onto the High Plains during the 1870s, stands of buffalo grass had receded westward about 150 miles. To buttress his argument, Wilber appealed to Scripture, interpreting Genesis 2:5–6 to mean that rain came only after God had created man to cultivate the soil. From Wilber's perspective, "[T]he Creator never imposed a perpetual desert upon the earth, but, on the contrary, has so endowed it that man, by the plow, can transform it, in any country, into farm areas."[5]

On a more mundane level, both Wilber and Aughey in 1880 outspokenly denounced the federal government's action to temporarily halt the sale of public lands as an attempt to stop all further settlement in the arid West. Actually, nothing was further from the truth because, since the inauguration of Rutherford B. Hayes in 1877, a group of reformers including Congressman Abram S. Hewitt of New York and John Wesley Powell, then director of the U.S. Geographical and Geological Survey of the Rocky Mountain Region, had sought to bring federal land policy in line with the nature of the arid West. That is the overall significance of the *Report on the Lands of the Arid Region*.[6]

Arguably the three most astute observers of the history and culture of the American West—all westerners themselves—agree on Powell's preeminence. Walter Prescott Webb called Powell's analysis of the land situation and the recommendations that followed "the most intelligent and comprehensive that had been made, and they have not been surpassed since." Bernard DeVoto described the *Report* as "one of the most remarkable books ever written by an American." He continued: "[I]t is a scientific prophecy and it has been fulfilled—experimentally proved.... It is a document as basic as the Federalist but it is a tragic document. For it was published in 1878 and if we could have acted on it in full, incalculable loss would have been prevented and the United States would be happier and wealthier than it is. We did not even make an effective effort to act on it till 1902." In that year, Congress authorized the establishment of the Bureau of Reclamation, of which DeVoto's friend, Wallace Stegner, called Powell the "spiritual father." Stegner later wrote the biography of Powell's career, *Beyond the Hundredth Meridian* (1954), and in 1962 he introduced and edited the republication of the *Report on the Lands of the Arid Region*.[7]

Let us be clear. Powell's primary interest was not horticulture but agriculture, in particular the development of irrigated farming. To the extent, however, that he shared the Jeffersonian ideal of yeoman farming and believed that every farm needed a vegetable garden for family use, his recommendations covered horticulture. So, too, did his proposals for collecting and storing vast reservoirs of water to allow for fruit growing (although not on the High Plains), ornamental horticulture, and, much later, urban development along the Front Range. Finally, by establishing that only a small portion of the arid West was susceptible to irrigation, Powell as surveyor left to the botanists and horticulturalists the task of finding and propagating plants adapted to the High Plains.

In his *Report*, Powell's immediate aim was to convince federal authorities to temporarily suspend settlement on unappropriated public lands in the West, whether through the numerous preemption, Homestead, Timber Culture, or Desert Land acts, but only until new laws could be enacted and new institutions established that would enable homesteaders to survive—indeed, thrive—in the arid West. He expressed the most concern for settlers who depended on waters from distant sources, in contrast to those living along small live streams such as those along the foothills of the Front Range.

Convinced that drought was the normal, not an abnormal, condition, Powell spoke out against those who promoted the agriculture of the humid East as practicable for the arid West. While many before him had criticized the promoters, Powell was the first who addressed "the problem as a whole, in the context of the entire West, and for the first time pointed out practical solutions for the difficulties of both the arid-lands homesteaders and the farmers in the sub-humid belt, where almost-adequate rainfall tempted men into a kind of farming a period of drought could destroy." In the sub-humid belt, Powell included the easternmost portion of the High Plains.[8]

The crux of the matter was water, namely, that there would never be enough to go around. To reiterate that water remains to this day more important than anything else in the arid West, I am reminded of an early discussion by our board of directors when we started the Wyoming Community Foundation. We asked the directors what they considered the most critical issue facing our state. Some hesitated through indecision and others through disagreement, until we got to Glyda May who, as a young child, had emigrated with her family from Iowa to homestead in Wyoming and, later, farmed and ranched with her husband near Wheatland on the eastern plains. Without giving the question a second thought, she decisively uttered "water!"

The solutions Powell proposed to address the scarcity of water were founded on an entirely novel concept of the relationship between land and water. To begin with, he devised a system to classify arid lands—those generally receiving less than twenty inches of precipitation annually—not according to conventional political boundaries but by their access to water: as irrigable, pasture, and timberlands. The timber category did not apply to the High Plains except for Powell's prophetic words (see Chapter 5) that forests be created artificially where precipitation was less than twenty inches, provided they received protection from rangeland fires.[9]

By irrigable lands, those that covered the smallest portion of the arid West, Powell meant lands dependent for water on relatively distant larger streams and not lands along smaller streams. While acknowledging that more study needed to be done, Powell used the experience of the California gold miners to estimate that 80 to 100 acres of irrigable land needed no more than a continuous flow of one cubic foot of water per second. Furthermore, by arguing that under such controlled conditions only the necessary amount of water would be used, he anticipated that "the waste [of water] now almost universal will be prevented."[10]

Powell envisioned that irrigation, the supply of water by "artificial methods," would take the risk out of farming in the arid West and thus make it more profitable than in humid regions. "Crops thus cultivated," he argued, "are not subject to the vicissitudes of rainfall; the farmer fears no droughts; his labors are seldom interrupted and his crops rarely injured by storms." In addition, the irrigable waters brought "fertilizing materials derived from the decaying vegetation and soils of the upper regions."[11]

Pasturage made up by far the greatest part of the arid West, containing vast areas of nutritious but scanty grasses and thus requiring farm units of no fewer than 2,560 acres—a far greater number than that stipulated by settlement under any existing public lands act. Most pertinent to horticulture,

Powell argued that those engaged in raising livestock "need small areas of irrigable lands for *gardens* and fields where agricultural products can be raised for their own consumption, and where a store of grain and hay may be raised for their herds when pressed by the severe storms by which the country is sometimes visited." In many cases, he continued, springs and brooks provided sufficient irrigation for those purposes, as well as for making farms "attractive and profitable."[12] Powell thus described the emerging operation known as the stock farm, combining pastoralism (ranching) and horticulture (agriculture)—in time, the most common form of farm on the High Plains.

After describing what he called the physical characteristics of the region, Powell concluded that making the most effective use of land and water required new forms of cooperation among settlers. In the case of irrigable lands, the diversion of waters from large, distant streams required cooperation and a large expenditure of both labor and capital. By way of precedent, Powell acknowledged "very successful" cooperative activities under ecclesiastical direction in Utah; outside Utah, he praised the "eminently successful" cooperative experience at Greeley. To enable such cooperation, in particular the construction of large upland reservoirs, Powell proposed that settlers of the public lands organize irrigation districts. He expected the settlers to establish their own rules and regulations for the use of water and division of lands. That was also true in the case of pasturage districts, "in which the residents should have the right to make their own regulations for the division of lands, the use of the water for irrigation and for watering the stock."[13]

Besides reflecting the Jeffersonian ideal of a nation of small farmers, Powell's proposal reflected his confidence in self-rule. Believing that by nature people yearn for self-rule, Powell argued that the institution of local districts would not only be effective in attracting permanent settlers but would also prevent monopolistic ownership of land and water and, what he most feared, the separation of title to land and to water.

In championing cooperation as the way to fairly ensure the distribution of scarce water, Powell advocated the establishment of communities, so that "the inhabitants of these districts may have the benefits of the local social organizations of civilization—as schools, churches, etc., and the benefits of cooperation in the construction of roads, bridges, and other local improvements." He believed grouping residences was practicable, even within pasturage districts, "by making the pasturage farms conform to topographic features in such manner as to give the greatest possible number of water fronts." Consequently, Powell rejected as impractical the conventional method of dividing lands into uniform tracts of square miles and townships and proposed a distinctly western method of surveying lands. Under his new method, a stream with enough water to irrigate 200 acres could supply water to 20 acres on each of ten farms, if they were properly located; under the old method, a uniform acreage owned by a single farmer could exclude nine other farms from irrigating.[14]

As a practical matter, then, water determines the value of land, which led Powell to conclude that "the right to use water should inhere in the land to be irrigated, and water rights should go with land titles." The right to water, furthermore, is secured by "priority of usage," or what was later known as the doctrine of prior appropriation.[15]

To fully appreciate the ramifications of this doctrine, which has underlain virtually all development—horticultural and otherwise—on the High Plains since the 1880s, let us contrast the doctrine of riparian rights that was in effect beforehand and why Powell found that doctrine entirely inapplicable to the arid West. The American legal scholar James Kent (1763–1847) wrote a classic definition of riparian rights, derived from the principles of English Common Law: "[E]very proprietor of lands on the banks of a river [*ripa* = Latin for stream bank] has naturally an equal right to the use of the water which flows in the stream adjacent to his lands, as it was wont to run without diminution or alteration." As a consequence of this doctrine, only owners of stream banks have the right to water. Each riparian must share equally with every other riparian and do nothing to prevent the others from getting their share. Owners of land away from streams have no rights to stream waters. Furthermore, water may be used while it runs over a riparian's land; however, it cannot be unreasonably detained or given another direction and must be returned to its natural channel—a situation that prevented the kind of irrigation Powell envisioned for the arid West.[16]

Indeed, for purposes of irrigation, Powell argued that water had no value in its natural channel; in most cases, moreover, water had to be diverted from its natural channel many miles from the lands where it was to be used and moved there through artificial canals. To cope with this situation, water rights needed to be severed from natural channels; at the same time, to prevent those rights from being controlled by the canal builders, water rights needed to be attached to the lands where the water was used. Thus, as Powell saw it, "the question for legislators to solve is to devise some practical means by which water rights may be distributed among individual farmers and water monopolies prevented."[17]

Toward that end, Powell prepared a draft bill "to authorize the organization of irrigation districts by homestead settlements upon the public lands requiring irrigation for agricultural purposes." Section 9 reads:

> That the right to the water necessary to the redemption of an irrigation farm shall inhere in the land from the time of the organization of the irrigation district, and in all subsequent conveyances the right to the water shall pass with the title to the land. But if after the lapse of five years from the date of the organization of the district the owner of any irrigation farm shall have failed to irrigate the whole or any part of the same, the right to the use of the necessary water to irrigate the unreclaimed lands shall thereupon lapse, and any subsequent right to water necessary for cultivation of said unreclaimed land shall be acquired only by priority of utilization.[18]

Not only had Powell provided the foundation for what Elwood Mead would draft as the "Wyoming Doctrine," connecting water rights to land titles, but Powell had also established the principle by which scarce water could be equitably distributed.

As suggested earlier, the doctrine of prior appropriation, which Powell had referred to as "priority of utilization," came out of the California mine camps where water was not only scarce but often had to be transported by channel for considerable distances. The doctrine applied to both riparian and nonriparian property owners. Essentially, it gave individuals the right to divert waters for beneficial use and gave the individual who made the earlier diversion a better claim to the water than an individual who made later use of that stream; thus, the maxim "first in time, first in right."

As a matter of practice, Colorado was the first state to wholly abrogate the doctrine of riparian rights, although it had taken some very specific steps during its territorial period. In 1876 Colorado was the first state to incorporate the doctrine of prior appropriation in its constitution (Article 16, Section 6), and in 1882 the justices of the Colorado Supreme Court (*Coffin v. Left Hand Ditch Co.*) clearly and decisively found that "the common law doctrine giving the riparian owner a right to the flow of water in its natural channel upon and over his lands . . . is inapplicable to Colorado. Imperative necessity, unknown to the countries which gave it birth, compels the recognition of another doctrine . . . [that] the first appropriator of water from a natural stream for a beneficial purpose has . . . a prior right thereto, to the extent of such appropriation."[19]

"Beneficial purpose" was left vague by the court; indeed, to this day no state statute has defined "beneficial use" except in the most general terms. The common meaning has been interpreted, however, to encompass irrigation for agriculture, water for livestock, and water for domestic and municipal use (including irrigating parks). Kansas statute would even include "irrigation of lawns and gardens" within the definition of domestic use.[20] Two important corollaries of the concept of beneficial use are, first, that water must not be wasted and, second, that if appropriated water is not used by an individual in a timely manner, that individual would lose the right to use it.

Thanks to the homework done by Greeley's Captain Boyd and his select committee, the Colorado Legislature in 1879 and 1881 established the first administrative process for the enforcement of the new water rights. As with the doctrine of prior appropriation itself, the administrative process stemmed from real experience with actual conditions, in this case, an argument between the irrigators of Fort Collins and Greeley over control of the waters of the Cache la Poudre River. The summer of 1874 had been very hot and dry, with insufficient water for all appropriators. Fort Collins farmers, higher up on the Cache la Poudre, diverted their water; Greeley farmers, left with too little water, objected on the grounds that they had prior rights. While this particular issue did not get resolved, the Greeley farmers—joined by farmers on the lower St. Vrain, Big Thompson, and, later, South Platte rivers— supported the select committee that drafted the water legislation. For purposes of water administration, provision was made to divide Colorado into water divisions and then into water districts that corresponded to areas irrigated by certain streams. Initially, only nine districts were defined, all in the South Platte watershed, to be known as Division 1. For each district, the governor appointed a water commissioner to assign priorities of appropriation as provided in written records. The process for obtaining such records was also established. To manage the administrative process, the legislature established the office of state engineer, whose major responsibility was to measure stream flow accurately—a particularly sensitive job, since most stream flow had been grossly overestimated by the appropriators.[21]

On the High Plains, the doctrine of prior appropriation was adopted by the Kansas Legislature in 1886 and by the Nebraska Legislature and the Dakota Territorial Legislature in 1889. The Wyoming Territorial Legislature adopted both the doctrine and the Colorado system of water-rights enforcement, as drafted by Elwood Mead and approved by voters as Article 8 of the Wyoming Constitution in November 1889. This article established that water was the property of the state; provided for a board of control composed of the state engineer and superintendents of the water divisions to supervise the appropriation, distribution, and diversion of the state's waters; and confirmed that "priority of appropriation for beneficial use shall give the better right."

Born on an Indiana farm, Elwood Mead (1858–1936) had moved on to the Colorado Agricultural College, where he helped horticulturist A. E. Blount and botanist James Cassidy establish the experiment station and, in 1885, had become the nation's first professor of irrigation engineering. In 1890, Mead left Fort Collins to become Wyoming's first state engineer. Like Major Powell, Mead believed irrigation would allow the arid West to be divided into small irrigated farms surrounded by vast grazing lands. Indeed, it was Mead, more than anyone else, who contributed to putting Powell's concepts into practice. In 1899, Agriculture Secretary James Wilson hired Mead to head the reestablished Division of Irrigation in the U.S. Department of Agriculture; and in 1924, after a long and distinguished international career, Interior Secretary Hubert Work appointed Mead to serve as commissioner of reclamation.[22]

While Mead's contributions to the development of horticulture were incidental to his primary interest in irrigation, from the land-grant colleges there emerged, around the same time, a concerted effort to apply research based on observation and experimentation to horticulture on the most arid lands. The leading proponent of that effort was Charles Bessey of Nebraska.

Notes

1. Josiah Gregg, *Commerce of the Prairies*, ed. Max L. Moorhead (Norman: University of Oklahoma Press, 1954), 362.

2. Samuel Aughey, *Sketches of the Physical Geography and Geology of Nebraska* (Omaha: Daily Republican Book & Job Office, 1880), 41–43.

3. Ibid., 44.

4. Charles Dana Wilber, *The Great Valleys and Prairies of Nebraska and the Northwest*, 3rd ed. (Omaha: Daily Republican, 1881), 74.

5. Ibid., 67, 70–71.

6. Henry Nash Smith, "Rain Follows the Plow: The Notion of Increased Rainfall for the Great Plains, 1844–1880," *Huntington Library Quarterly* 10 (1947): 181–182, 185.

7. Walter Prescott Webb, *The Great Plains* (1931; repr. Lincoln: University of Nebraska Press, 1981), 422; Bernard DeVoto quoted in Wallace Stegner, *Beyond the Hundredth Meridian: John Wesley Powell and the Second Opening of the West* (1953; repr. Lincoln: University of Nebraska Press, 1982), xxii; John Wesley Powell, *Report on the Lands of the Arid Region of the United States*, ed. Wallace Stegner (Cambridge: Harvard University Press, 1962), xxiii, 23.

8. Powell, *Report*, xi, 59.

9. Ibid., 25.

10. Ibid., 18.

11. Ibid., 20, 96.

12. Ibid., 31–32, 34–35.

13. Ibid., 21–23, 40, 42–45.

14. Ibid., 33–34.

15. Ibid., 54–55.

16. James Kent, *Commentaries on American Law* (New York: Clayton, 1840) 3: 438–439; Robert G. Dunbar, *Forging New Rights in Western Waters* (Lincoln: University of Nebraska Press, 1983), 59–60.

17. Powell, *Report*, 53–55.

18. Ibid., 44–45.

19. Robert G. Dunbar, "History of Agriculture," in Leroy Hafen, ed., *Colorado and Its People* (New York: Lewis Historical Publishing, 1948) 2: 125–126.

20. Frank J. Trelease, "The Concept of Reasonable Beneficial Use in the Law of Surface Streams," *Wyoming Law Journal* 12 (Fall 1957) 12: 7–12.

21. David Boyd, *A History: Greeley and the Union Colony* (Greeley: Greeley Tribune Press, 1890), 122–126; Dunbar, "History of Agriculture," 127–128; Dunbar, *Forging New Rights*, 86–98.

22. James R. Kluger, *Turning on Water with a Shovel: The Career of Elwood Mead* (Albuquerque: University of New Mexico Press, 1992), 8–26.

5

Science and Its Application to Horticulture

The year 1887 marked a watershed in the history of horticulture on the High Plains. By then, the idea that horticulture contributes to better living had supporters both on the homestead and in the community. Legislation promoting horticulture, including the establishment of state horticultural societies, had passed in Nebraska, Kansas, and Colorado. New laws and regulations concerning water and land, thus horticulture, had been decreed in these states as well as in Dakota and Wyoming territories. And while federal land policy had yet to be reconciled with the nature of the arid West, an overall development plan had been suggested and its obstacles defined.

Before 1887, horticulture, generally speaking, had advanced by trial and error, in an unsystematic and fragmented manner. After 1887, horticulture would advance by the application of science, backed by federal financial and administrative support. Furthermore, for the quarter century ending in 1914, horticulture on the High Plains would be guided largely by the work of three professors: Charles Bessey of Nebraska, Aven Nelson of Wyoming, and Niels Hansen of South Dakota.

While Major Powell had been interested primarily in the development of irrigated farming on a large scale, these three professors devoted their research, teaching, and service to what grew and what could grow on the land Powell had classified as pasturage, which included most of the High Plains. Of the three, Bessey was the earliest proponent of applying the principles of science to the practice of agriculture and horticulture. In fact, it was he who had coined "Science with Practice" as the motto for the Iowa State College of Agriculture and Mechanic Arts, now Iowa State University, before he moved on to the University of Nebraska.

For scientific principles to be applied to agricultural practices, Bessey argued for more and better research at the land-grant colleges. Toward that end, he was among the earliest advocates of federal funding for agricultural research. It is unknown whether, while still teaching at Iowa, he attended a conference on the topic in Washington, D.C., called by the United States commissioner of agriculture, George B. Loring. We do know that immediately following the meeting, Bessey was called by his colleague Seaman Knapp, professor of agriculture and soon to be president of Iowa State, to help draft a bill providing for federal aid to a national network of agricultural experiment stations. Introduced in May 1882 by Representative Cyrus C. Carpenter of Iowa, the bill did not pass out of committee; however, after another five years of negotiations, Senator William H. Hatch of Missouri succeeded in getting the legislation passed and signed into law by President Grover Cleveland.[1]

The act of March 2, 1887, known as the Hatch Act, inaugurated perhaps the most successful partnership in U.S. history between the federal government and the states: it created a nationwide system of agricultural experiment stations, with each station operated separately by its respective land-grant institution. The act's overall purpose was "to aid in acquiring and diffusing among the people of the United States useful and practical information on subjects connected with agriculture, and to promote scientific investigation and experiment respecting the principles and applications of agricultural science." Such language was not new; it reflected a traditional outlook going back to America's founders and resembled, for example, the enabling language of the American Philosophical Society, founded in Philadelphia in 1743 for the purpose of "promoting useful knowledge." Furthermore, the language of the Hatch Act distinguished not so much between "pure" and "applied" science as between science and its applications, the latter distinction a favorite topic of Charles Bessey's.

Indeed, given Knapp's generally nonscholarly interests, one can reasonably assume that Bessey drafted Section 2 of the act, which set forth the scientific research and experimentation to be conducted by the stations and read in part: "that it shall be the object and duty of said experiment stations to conduct original researches or verify experiments on the physiology of plants and animals; the diseases to which they are severally subject, with the remedies for the same; the chemical composition of useful plants at their different stages of growth ... the capacity of new plants or trees for acclimation ... [and] the adaptation and value of grasses and forage plants."[2] By the time the Hatch Act became law, Bessey was professor of botany and horticulture and dean of the Industrial College (including agriculture) at the University of Nebraska.

Born to well-educated parents and reared on a farm in north-central Ohio, Charles Edwin Bessey (1845–1915) attended rural primary school and secondary academy. He taught school for about two years, not unusual in those days, before entering Michigan Agricultural College, a land-grant college that is now Michigan State University. He originally intended to study civil engineering but switched to botany at the suggestion of Albert N. Prentiss, instructor in botany and horticulture, and Theophilus Abbot, professor of English literature and college president. Bessey's course of study included the classics and the sciences; he earned a Bachelor of Science degree in November 1869. After graduation he remained at Michigan Agricultural College, working as greenhouse assistant to George Thurber, later editor of the *American Agriculturalist*. In the fall of 1870, Bessey began his college career as instructor of botany and practical horticulture at Iowa State College. Because the college was new and Bessey arrived with horticultural experience, he was put in charge of landscaping the campus and starting the college garden. In honor of that pioneer work, two years later his alma mater awarded him an honorary Master of Science degree. Bessey spent three long winter vacations (1872, 1873, and 1875) at Harvard under the direction of America's principal botanist, Asa Gray. Considering Bessey's future preeminence, it might seem unusual that he did not earn a Ph.D. in botany; however, the tradition of college faculty with doctoral degrees did not begin until the twentieth century.[3]

By the time Bessey arrived at Nebraska, he had adopted the view that the usefulness of a college education to a farmer or horticulturist was not in teaching students how to cultivate the soil but in helping them understand nature's laws—the basic principles of science—so they could apply those principles intelligently to farm or garden. In his inaugural address to faculty and students of the Industrial College, Bessey amplified his view: "The College can tell how plants grow, but it is not within its province to tell what plants to grow. The former is general and true for all places and under all conditions; the latter is local and is modified by a thousand contingencies." College faculty, he continued, cannot hope to "compete in the discovery of new methods of practices with the thousands of quick-witted gardeners whose livelihood and financial prosperity depend upon the discovery of better methods."[4]

Two years later, while informing the Nebraska Horticultural Society about his plans to establish a Horticulture Department within the Industrial College—the first such academic department on the High Plains—Bessey reiterated his view on science and its applications. Horticulture can only be pursued with a solid knowledge of modern botany: "Every good horticulturalist is to a certain extent a botanist in the better sense of that term." Thus, in the course of study of horticulture, "the student must begin with a mastery of much of what is now included in the science of botany. We have, therefore, placed before the student in horticulture, first, the study of the general anatomy of plants. He is to make himself familiar with the structure of roots, stems, leaves, flowers, fruits, etc., by the actual study of the things themselves, aided by a good microscope and such other appliances as may be necessary."[5]

Committed to the empirical (observable and experimental) approach to instruction, Bessey pioneered the use of the microscope in the Nebraska classroom. Shortly before he officially took up his Nebraska duties, he had obtained from the university's regents a special appropriation to purchase the first six microscopes at Nebraska and other laboratory materials. During his first summer, he started a botanical garden; from correspondents statewide and from the United States Department of Agriculture (USDA), he received enough specimens to establish an herbarium. Bessey distinguished between laboratory learning and the experience of working in a nursery, greenhouse, or orchard. Only at the University of Nebraska could a young Nebraskan learn scientific horticulture, including experimental work in such areas as fertilization, propagation, hybridization, and plant diseases. As to vocational training—grafting, budding, pruning, orchard culture, nursery management—that should be left to members of the Horticultural Society.[6]

By the time Bessey reached Nebraska, he was also well aware of the criticism of the course of study separately established for agricultural students, and he argued that the reputation of the industrial or agricultural college would improve greatly if the college stuck to teaching how plants and animals grow and left the "contingencies" to others. This was no expression of disdain for vocational instruction or agricultural practitioners. Instead, it reflected Bessey's view that everyone needs to know how to read and write,

Professor Bessey's botanical laboratory. Courtesy, Archives and Special Collections, University of Nebraska, Lincoln

know about the general geography and history of the world, and know about plants and animals. In a democracy, furthermore, "it is a grave mistake for any people to separate its classes in their education, for such separation tends to the establishment of castes and clans, and prohibits that ready change from class to class which is so desirable in a free and intelligent community." Years later, Bessey wrote on the same theme: "[T]he longer I teach, the more I am impressed with the value of training all classes of people together." In other words, whether training to be a horticulturist or a physician, every student should receive the same basic instruction and thereby be prepared for citizenship in the modern world.[7]

Still in Iowa, Bessey had recognized that key to first-rate college instruction was improving the quality of college preparatory instruction in the sciences. Toward that end, in 1880 he had published a botanical text, primarily for high school students, that went through numerous editions during his tenure at Nebraska. In 1892, he published an exercise book for Nebraska teachers with no prior botanical knowledge, to be used in introducing students to the study of nature in their schools' surroundings. Bessey also contributed to the writing of state standards for teaching high school botany, issued in 1897; prepared the syllabus for a one-year high school course in botany, complete with microscopes, dissecting sets, and other aids; and recommended which plants should be selected for laboratory study. In 1901, he drafted legislation providing for the teaching of agriculture and horticulture in rural and normal (teacher-training) schools. To the Board of Agriculture, he expressed the hope "that the day will come when there will be something of formal instruction in the elements and rudiments of agriculture and horticulture in every school in the great state of Nebraska."[8]

Beyond the formality of public instruction, however, Bessey welcomed letters concerning nature from young people. Such letters, he told the *Nebraska Farmer*, "bring back pictures of my own boyhood, spent on my father's farm in Ohio. I remember the longing I had for information as to the nature of many of the things about me, and I have no reason for thinking that the boy[s] of today in Nebraska are less curious than boys twenty-five or thirty years ago in Ohio. So let the boys and girls of the farms of today send in their questions for it will indeed give me pleasure to discuss them." They did so throughout his tenure at the University of Nebraska. When, for example, an eighth grader from Beatrice wrote in search of literature that might help him prepare for an essay competition on the topic of the "ten best trees to grow in Nebraska, and how to plant and cultivate them," Bessey responded with bibliographical suggestions and a note: "[I]f I can help you further, be sure to write me again."[9]

For adults outside the classroom, Bessey used his various public positions to illustrate the application of botany to horticulture. During his first year at Nebraska, he lectured the State Horticultural Society on "Cryptogams"—the term then used for non-flowering plants such as ferns, mosses, and fungi—specifically, on those fungi injurious to vegetables. In his second year (1885), the State Board of Agriculture appointed him state botanist, a position he held until 1907. His duties included preparing an annual report, writing lectures on grasses and forage plants that were published by the board, and preparing horticultural exhibits at state fairs. As an officer of the board, the state botanist received reimbursement of expenses.[10]

At his first meeting with the university regents, Bessey indicated his intention to engage his college faculty in scientific experiments on the effect of temperature and humidity on soil fertility. During his first full year on the job, Bessey wrote short articles on diseases of apples and plums. After passage of the Hatch Act in 1887, for which he had enlisted the endorsement of the State Horticultural Society, Bessey obtained regent approval to use federal funds in support of the research he had earlier proposed. Moreover, Bessey was appointed, and served for two years, as the first director of the Nebraska Agricultural Experiment Station.[11]

With the backing of Robert Furnas, longtime horticulturist and former governor, Bessey put the Farmers' Institutes on solid financial and organizational footing; he lectured frequently himself. In addition to his speaking engagements, Bessey contributed articles to both the general and farm press of Nebraska. As acting chancellor (June 1888–June 1891), all the while remaining college dean and experiment station director, Bessey instituted the tradition of annually inviting a delegation, consisting of members of the State Board of Agriculture and the State Horticultural Society, to visit the agricultural experiment station and publicly report on its findings.[12] Externally, such visits served as a vehicle for publicizing the university's role in promoting useful knowledge; internally, it pressured faculty and students to work hard and to produce within the context of the land-grant college mission.

As very much an economic botanist settling in a state botanically unexplored, Bessey's first major research project was to catalog the state's plants. Once classified, Bessey and his students could turn to considering which native plants might be practical for cultivation; the impact of non-native plants, especially noxious weeds, on the local flora; and what imported plants could be successfully cultivated in Nebraska. Because of the state's geography, grasses and forage plants were of primary interest; in a contrary sort of way, so too was the lack of trees, especially on the High Plains in western Nebraska.

Before Bessey, the only attempt at describing Nebraska flora was a plant list published in 1875 by Samuel Aughey. Since that list was generally considered faulty and incomplete, Bessey started anew, relying primarily on the collecting by his students and, to a lesser extent, on his correspondents and his own fieldwork. Like Aven Nelson later in Wyoming, Bessey had drawn a devoted following of students attracted to fieldwork. Beginning in 1886, two years after Bessey's arrival at Nebraska, his best students associated themselves into an informal academic fraternity, the Botanical Seminar, known affectionately as "Sem Bot." Out of this group came the handful of students and graduates who undertook the botanical survey of Nebraska.

The plan was to publish the "Flora of Nebraska" in twenty-five sections of roughly fifty pages each. Although endorsed by the University of Nebraska, the State Board of Agriculture, and the State Horticultural Society, sufficient funding did not materialize, so only two sections were published by 1895. Meanwhile, Bessey's student, Herbert J. Webber, had helped prepare a catalog of Nebraska grasses and forage plants in 1890. Among other early students,

J. M. Bates, Jared Smith, and Albert Woods collected grasses on the High Plains of western Nebraska, while Webber, Frederick Clements, and Herbert Marsland did general collecting in the central Sandhills. Thomas A. Williams collected in northwestern Nebraska, and Per Axel Rydberg collected in southwestern Nebraska. Except for Rydberg, who fulfilled his career at the New York Botanical Garden, the remaining students would take positions with the United States Department of Agriculture.[13]

During the 1880s, Bessey had done his own botanical fieldwork, most notably in the winding canyon of Lone Pine Creek, which runs from a point about twenty-five miles south of Ainsworth, on the northeast edge of the Sandhills, north into the Niobrara River. There he reported finding "a blending of the Eastern and Western floras in a most unusual way." Of primary interest, he found an abundance of what he called Rocky Mountain pine (*Pinus ponderosa* Laws.), allowing him to extend its range eastward from the Black Hills—thereby updating its geographic distribution as first described by Charles Sprague Sargent, the leading authority on trees in America. In Lone Pine Creek Canyon, Bessey also identified the black walnut (*Juglans nigra* L.) and extended westward Sargent's distribution for that species as well. "I doubt whether there is any place on the continent," Bessey later told the Botanical Club of the American Association for the Advancement of Science, "where the black walnut and the Rocky Mountain pine grow normally side by side." Not long afterward, Bessey put forward the hypothesis that eastern species of trees and shrubs were moving westward and that western species were slowly retreating. To support his notion, he had observed that "so few of the western trees and shrubs have come down the streams, especially as prevailing winds are also from the westerly parts toward the east."[14]

Lest one imagine that Bessey was all "pure" science, he also told his fellow botanists in New York City that, when possible on his searches, he watched for edible plants. "At a home [near Lone Pine Creek] where I stopped for dinner," he reported, "I was treated to choke-cherry pie, which was very palatable indeed."[15]

Bessey's hosts along Lone Pine Creek probably counted among his many correspondents statewide who collected plant specimens and mailed them to him in Lincoln. Sometimes, Bessey took the lead by asking his correspondents to find particular plants. In response, for example, a rancher in western Nebraska apologized that it was too late in the season to supply Bessey with good samples of buffaloberry fruit (*Shepherdia argentea* (Pursh) Nutt.). Since buffaloberry made fine jellies, its fruit had already been gathered for canning by the homesteaders.[16]

Despite his chores in academic administration, during the 1890s Bessey found time to give several lectures on the botany of fruit trees and small fruits to the State Horticultural Society. His fundamental advice was always the same, namely, that to grow plants successfully, horticulturists needed to go beyond the trial-and-error methods of the past and apply their understanding of nature's laws—the basic principles of science—to the cultivation of their orchards and gardens. In describing the outward characteristics of plant diseases, he covered not only crops such as wheat and barley but also common farm vegetables such as lettuces, tomatoes, squashes, and potatoes.[17]

As state botanist, Bessey's publications were necessarily oriented more toward agriculture than to horticulture. At the same time, he issued numerous warnings on the rapid spread of unwanted plants or weeds that affected both. Although he had identified at least 125 native plants as weeds, he was most concerned about the introduction of non-natives, in particular the tumbleweed or Russian-thistle (*Salsola kali ssp. tragus* (L.) Aellen). Although not a true thistle, Russian-thistle had been introduced inadvertently with flaxseed from Eastern Europe to South Dakota in the mid-1870s, and from there it spread rapidly throughout the High Plains.

Bessey prepared an agricultural experiment station bulletin on the Russian-thistle, saw to preparation of an exhibit of its characteristics at the Nebraska State Fair, and spoke widely on the subject, believing the first step toward its control and eventual eradication was correct identification. He did not limit himself, however, to simply identifying the Russian-thistle. Instead, he was among the earliest of his colleagues at the land-grant colleges to actively campaign for legislation requiring landowners to eradicate this noxious weed. While he favored adoption of the tough Wisconsin Weed Law of 1885, he settled for an amendment to existing Nebraska law against the Canada thistle (*Cirsium arvense* L.) to include the Russian-thistle:

> Every owner or possessor of land shall cut or mow down all Canada thistles growing thereon or in the highway adjoining the same, so often as to prevent their going to seed, and if any owner or possessor of land knowingly shall suffer any such thistles to grow thereon or in any highway adjoining the same, and the seed to ripe so as to cause and endanger the spreading thereof, he shall forfeit and pay a fine not less than ten dollars and not more than forty dollars; and any person may enter upon the land of another, who shall neglect or refuse to cut or mow down such thistles, for the purpose of cutting or mowing down the same, and shall not be liable to be sued in an action for trespass therefore.[18]

Despite the good intentions of the amended law, anyone who drives across western Nebraska, or anywhere else on the High Plains, today and glances over to the fence line will recognize that the attempt to eradicate the Russian-thistle has been a failure.

Bessey devoted much of his teaching, research, and public service to Nebraska's most common and economically significant plants. His scientific curiosity, moreover, led him to consider the absence of forests in Nebraska. To be sure, consideration of the treeless prairies, and whether more trees would increase precipitation, had been "in the air" for over a decade. The fact that Bessey headed the Forestry Committee of the State Horticultural Society soon after his arrival in Nebraska suggests that his interest in trees was not new. Specifically, he sought an answer as to why the Sandhills were virtually treeless and whether that condition had always been the case. His hypothesis, that the Sandhills had once been forested, would form the basis for a series of "re-forestation" experiments that culminated in 1902 with the establishment of the first artificial national forest near Halsey in the western Sandhills.

Again, while forestry by itself is not part of horticulture, the results of Bessey's work and the fact that the Bessey Nursery of the Nebraska National Forest served for years as the most important source of seedlings for windbreaks on the High Plains place this particular development within the boundaries of our story.

To fully appreciate Bessey's contribution to tree planting and thereby to making the High Plains more livable, it would be well to fix in the reader's mind the nature of that truly unique region known as the Sandhills. Sparsely populated, it encompasses around 18,000 square miles of north-central Nebraska, from the 98th to the 103rd meridian (Grand Island to Alliance) and from the 41st to the 43rd parallel (North Platte to the South Dakota line). Rising in elevation from 1,800 feet in the east to 4,000 feet in the west, the topography consists of rolling hills—actually, treeless sand dunes covered primarily by grama grasses (*Bouteloua* species) and interspersed with small lakes.

In describing the Sandhills, Bessey's star student, Per Axel Rydberg, emphasized the action of the ever-present wind: "The Sandhills change their configuration constantly. Whenever the sand is not held together by the roots of plants or by moisture, or is not otherwise protected, it is little by little carried away by the wind. If a spot on a dry hill becomes bare and the loose sand is blown away, a small hollow is made, the surrounding grass dies from drought, the dry sand, no longer held together by the roots, slides down into the hollow and in its turn is borne away, and thus the hollow becomes gradually larger and larger."[19] Those hollows, known as "blowouts," can be seen throughout the Sandhills.

In 1891, the year Bessey published the first "Report upon the Native Trees and Shrubs of Nebraska," he persuaded Bernhard E. Fernow, chief of the Division of Forestry at the USDA, to provide the seedlings and funding for an experimental conifer test plot in the Sandhills. Fernow, in turn, asked Bessey to find suitable land and labor at no cost to the government, which Bessey did despite being "considerably provoked" by the request because of his many administrative duties (he still served as the university's acting chancellor). Luckily, Bessey quickly discovered that his colleague in entomology, Lawrence Bruner, and Bruner's brother owned acreage in Holt County in the northeastern Sandhills that they willingly dedicated to the experiment.

Under Fernow's general instructions, Bruner's brother took charge of planting 20,000 conifer seedlings: ponderosa pine (*Pinus ponderosa* Laws.), the non-native jack pine (*Pinus banksiana* Lamb.), Scotch pine (*Pinus sylvestris* L.), and Austrian pine (*Pinus nigra* Arnold). As "nurse" or shelter trees, Bruner *frère* planted black locust (*Robinia pseudoacacia* L.), boxelder (*Acer negundo* L.), hackberry (*Celtis occidentalis* L.), and black cherry (*Prunus serotina* Ehrh.). Fernow had directed that some seedlings be planted on cultivated land and on virgin land broken only by the furrows created as the seedlings were set out. Bruner *frère* did no further cultivation, but he did take steps to ensure protection of the test plot against grazing livestock and range fires.

By the end of the first growing season, fall 1891, roughly 5 percent of the seedlings on cultivated land were alive; the remainder had succumbed to the wind, which had taken away the sandy soil and left only "blowouts." On the uncultivated land, more than 50 percent of the seedlings had survived, with jack pine faring the best, closely followed by ponderosa pine. Subsequent to that first season, the Holt County experimental plot dropped out of sight, at least from Lincoln, and no further reports were made. "We supposed, as probably did everybody else who knew of the original planting," Bessey later wrote, "that the trees had disappeared and that we had simply one more case of the wreck of tree planting such as were familiar to us in the days of the forest homesteads."[20]

In republishing his report on Nebraska trees and shrubs through the State Horticultural Society, Bessey added the request that anyone finding errors in the text notify him, and he invited readers throughout Nebraska to collect and send him specimens. In response, Bessey received correspondence about the tree-planting experiment of J. C. Toliver, district judge and amateur horticulturist at Ainsworth. Believing the subsurface water level throughout the

Sandhills was sufficiently high to ensure germination, Toliver had collected ponderosa pinecones and placed them in buckets of water to separate fertile seeds, which he planted on his Timber Culture Act claim. While his experiment in planting conifers from seed failed, Toliver remained convinced that the Sandhills were "the natural home of the pine tree" and urged Bessey and the university faculty to pursue research on reforestation.[21]

In using the term "reforestation," Toliver clearly assumed that the Sandhills had once been covered with trees, a belief supported by anecdotal evidence from fellow homesteaders who had found ancient tree trunks along watercourses. By forestry, he meant what is sometimes called tree farming, namely, growing trees primarily for wood, not for windbreaks or shade. Within that context, Bessey reported to the State Board of Agriculture in 1893, based on what he knew about the Sandhills, that "it is possible to cover great tracts of this country with trees and shrubs, from which a good revenue might eventually be derived." He expressed the hope that the Nebraska Legislature, the deliberative body that gave America the idea of Arbor Day and had designated Nebraska the "Tree Planter State," would provide incentive to start the enormous enterprise of reforesting the plains. Indeed, he proposed as his "favorite idea" that the state acquire large tracts of land for the establishment of forest reserves.[22]

Seven years later, in late 1900, a grand opportunity came when Bessey received notice from William L. Hall, superintendent of the Tree Planting Section of the USDA Bureau of Forestry, that he planned to lead a group of foresters in investigating the practicability of setting aside public land in the Sandhills for a major forest experiment. During the spring and summer of 1901, Hall made Bessey's departmental office his field headquarters. Bessey recalled that one day Hall announced that he had learned enough about the Holt County planting ten years earlier to warrant an inspection trip. Bessey noted:

> I confess to have been quite troubled over the fact that Mr. Hall was to visit this plantation, as I felt sure that it must have disappeared and its disappearance would be an argument against the possibility of foresting the Sandhills in spite of any carelessness that might have resulted in the failure of the experiment. So I waited for a week or ten days in a more or less troubled state of mind, when one day Mr. Hall walked into my office in a state of great excitement. I called to him and said "what is the matter, Mr. Hall?" When he answered, "why, I have seen them," I said, "seen what?" He said, "Those trees." I said "What trees?" "Oh, those trees planted in Holt County ten years ago." Hall had seen conifers eighteen to

twenty feet high formed into a dense thicket in which forest conditions had appeared; and with greater growth than on similar trees planted in eastern Nebraska.[23]

Hall's trip to Holt County helped dispel doubt about the viability of growing trees in the Sandhills; however, Holt County was not the primary geographic area of interest to the Sandhills Reconnaissance Survey. Hall's team had concentrated its work in western Nebraska from Kearney to North Platte, then northwestward along the North Platte to the Wyoming line. The team included two young foresters who would distinguish themselves on the High Plains: Royal S. Kellogg would conduct the earliest inventory of western Kansas plants, useful to the establishment of the experiment station at Hays; Charles A. Scott would serve as first superintendent of the Nebraska National Forest, then as state extension forester of Colorado, and would complete his career as director of the New Deal's Prairie States Shelter Belt Project in his native Kansas.

For a short time, Hall's superior, Gifford Pinchot, chief of the Forestry Division, joined the survey. Notified in advance of the visit, Bessey arranged for the State Horticultural Society to hold its summer meeting in Kearney and thereby to hear Pinchot, Frederick H. Newell of the Geological Survey, and other federal officials.

The timing of Pinchot's visit, and Hall's subsequent recommendations for the reforestation of the Sandhills, was fortuitous. On September 15, 1901, Theodore Roosevelt succeeded William McKinley as president of the United States; Pinchot, Roosevelt's longtime trusted adviser, gained influence over matters of conservation and natural resources. At Hall's request, meanwhile, Bessey had provided Governor Ezra P. Savage with suggested items to be included in a letter to the president and then wrote his own letter directly to Roosevelt, emphasizing the economic potential of reforestation in the Sandhills. Bessey understood Pinchot's position on managing public forests primarily for timber production. In fact, the Roosevelt administration introduced the public policy that recognized timbering and grazing on public lands as forms of agriculture.[24]

On April 16, 1902, President Roosevelt issued the proclamation that established the Nebraska Forest Reserves in the Sandhills. He specified two areas: the Dismal River Forest Reserve, 85,000 acres between the Middle Loup and Dismal rivers southwest of Halsey (just west of the 100th meridian) and the Niobrara Forest Reserve, 123,000 acres between the Niobrara and Snake rivers south of Nenzel (101st meridian). A third area, the North Platte

Early plant planting crew, Bessey Nursery. Courtesy, U.S. Forest Service.

Division, 347,000 acres south of Hyannis, was established by the president in 1906 but was retracted and opened for homestead entry under the Kinkaid Act of 1904.

During that first summer of 1902, the Forestry Division cleared land near the Middle Loup and prepared seedbeds for the nation's first federal tree nursery, known to this day as the Bessey Nursery, with the capacity to grow 4.5 million seedlings. The first beds were sown with ponderosa seeds collected in the Pine Ridge area of northwestern Nebraska and the Black Hills of southwestern South Dakota.

The obstacles to growing large quantities of stock at the Bessey Nursery were many, and the early results were discouraging. One of Bessey's graduate students, Raymond Pool, later reminisced that the fine mineral soil was far from ideal:

> It lacked physical stability and its original fertility was low. Fertilizers, both inorganic and organic, were thoroughly mixed with the soil. Wind guards were erected to protect the delicate seedlings. An irrigation system was installed in order to provide a constant supply of water for favorable growth

First Bessey Nursery office. Courtesy, U.S. Forest Service.

conditions in the seed beds. The long beds of seedlings were shaded, in the beginning, by a canopy of wooden slats that were stretched over a framework attached to cedar posts about seven feet high. This canopy afforded the necessary half shade for the seedlings and it was high enough to enable the workmen to move about with ease beneath its slatted top.[25]

A similar canopy would be constructed for the Cheyenne Horticultural Field Station in the early 1930s.

One of the most discouraging early setbacks occurred after tens of thousands of seeds had been planted, seedbeds well cultivated, and seedlings reached about two inches in height. Without warning, a fungal parasite attacked the seedlings, causing them quickly to wilt, collapse, and die. It was left to Carl Hartley, a Nebraska native who had earned his master's degree in botany at Nebraska and worked as a forest pathologist for the USDA, to study the situation at Bessey Nursery. In time, he developed a technique of seedbed sanitation and a chemical treatment of the soil that prevented the process known as "damping off," in which disease organisms cause early seedling death.

As part of starting the Dismal River Forest Reserve, federal workers planted about 100,000 jack pine seedlings from Minnesota in the spring of

Plowing planting furrows for seedlings, Bessey Nursery. Courtesy, U.S. Forest Service.

1903, followed by planting ponderosa pine seedlings from the Black Hills. Fewer than 20 percent survived beyond three years. After all attempts at broadcasting the seeds of ponderosa and jack pines, eastern red cedar (a Nebraska native, *Juniperus virginiana* L.), and Colorado blue spruce (*Picea pungens* Engelm.) on the hills near the nursery failed completely, foresters turned to planting seeds in the nursery seedbeds, transferring year-old seedlings into "transplant" beds for another year or two, then setting out seedlings in permanent locations. The first ponderosa pine were set out in 1904, and within a very few years the most promising species proved to be the two Nebraska natives—ponderosa pine and eastern red cedar—and jack pine.[26]

While the Nebraska forest reserves and attached tree nursery may have been Charles Bessey's most visible legacy to the development of horticulture on the High Plains, his observations about the limits of horticulture in the Sandhills and western Nebraska were equally noteworthy. Bessey knew that irrigation had been relatively primitive in technique and limited to very small acreages, mostly confined to the North Platte and its tributaries such as Lodgepole Creek, and he appreciated the life-saving role of the windmills noted in Chapter 3.

Bessey knew from Rydberg that crop farming in western Nebraska was marginal at best. While botanizing west of the 100th meridian in the summer

Seedling cultivation, Bessey Nursery. Courtesy, U.S. Forest Service.

of 1893, Rydberg had seen fine products of small farms—potatoes, cabbages, tomatoes, onions, watermelons, and cucumbers. Generally, however, Rydberg noted that farming did not fare very well, which he ascribed to the hot, desiccating winds of August coming off the Sandhills rather than to the lack of rainfall.[27]

Reflecting on the drought of the 1890s and the resulting depopulation of western Nebraska, Bessey expressed the view to the State Board of Agriculture that "instead of lauding this western region as the best in the world for the growing of corn, wheat, oats, and other farm crops, it would have been much better [on the part of early promoters] to have held it up as one of the most promising for the growing of herds of cattle, horses, and sheep." To be sure, the land could support good small farms; however, crop farming, which destroyed native grasses, was essentially uneconomical because it destroyed rather than took advantage of native vegetation.[28]

By no means, however, did Bessey condone the overgrazing that had contributed to the deterioration of so much land. His criticism of "boomer" publicity about the opportunities for dry-land farming, combined with his own judgment as to the limits of farming in western Nebraska, help explain his

Shade frame, Bessey Nursery. Courtesy, U.S. Forest Service.

doubts about the Kinkaid Act of 1904—authored by Congressman Moses P. Kinkaid of O'Neill (eastern edge of the Sandhills) and signed by President Roosevelt—amending homestead laws to increase entries from 160 to 640 acres on arid (non-irrigable) lands. Meant exclusively for the thirty-seven Nebraska counties west of the 98th meridian, the act aimed to stem emigration and encourage small farms. Subsequent federal appropriations specified free distribution of seedlings from the Bessey Nursery to homesteaders in the area covered by the Kinkaid Act.[29]

While Bessey's retirement as state botanist in 1907 marked the end of his official public career, he continued as teacher and scholar, not neglecting his role as missionary for science and its applications. He reflected on that relationship in welcoming the State Horticultural Society, which held its annual meeting that year at the Botany Department in Nebraska Hall. In the twenty-one years faculty and orchardists had known each other, he reminisced, "we of the Department have learned to have a much higher regard for your knowledge of the great art of growing fine fruits, while on the other hand I trust that you have learned that the science of botany contains much knowledge that has proved useful to you in your work." Bessey happily acknowledged that scientists and practitioners had not only come to understand each other better but also to work together more closely for the good of Nebraska.[30]

Bessey fervently believed that being a land-grant college faculty member entailed a special responsibility to and engagement with the citizens of one's state. At a banquet held in his honor in Lincoln in 1913, attended by representatives of twenty statewide organizations connected to agriculture and horticulture, Governor John H. Morehead noted that Bessey's name was perhaps the most widely recognized by Nebraskans; Chancellor Avery observed that Bessey had come into more contact with the people of Nebraska than anyone else connected with the university.

Beneath the official expressions of appreciation lay real affection and admiration for Bessey as a person. When still teaching high school in the Sandhills region, State Superintendent of Public Instruction James E. Delzell reminisced that he had sent Bessey a plant for identification. Bessey reported back the description and ended his note with the observation that "[i]t's a very pretty little flower." Delzell added: "You don't know how that last [sentence] impressed me with the kindly thought he had for that little flower [species not mentioned]. I thought surely that man has a kindly thought for all the students that are with him."[31]

Expressions of esteem for Bessey also came from S. C. Bassett, master of ceremonies, a year older than Bessey and one of his longtime correspondents and friends: "None know him but to love him, none name him but to praise [him]." Bessey responded by reiterating his conviction that the future of Nebraska depended on the strength of the university in providing thorough instruction not simply to young people but to all the inhabitants of the state.[32]

Beyond Nebraska, Bessey was looked upon as the dean of natural sciences by his botanical and horticultural counterparts Carruth and Kellerman in Kansas, Cassidy and Crandall in Colorado, Hansen in South Dakota, and Nelson in Wyoming. Hansen and Nelson viewed Bessey as their mentor—Hansen had studied under Bessey at Iowa and kept in touch with him to the end of Bessey's life, and Nelson admired Bessey for both his scholarly and public service work. As so often happens with good people in mediocre institutions, Nelson sought Bessey's outside advice on how to cope with the obstacles to progress in Wyoming and, on occasion, even discreetly asked Bessey about job opportunities elsewhere. It is fair to say that of all the High Plains states, Nelson's Wyoming had the furthest to go in developing its agriculture and horticulture.

Notes

1. Joseph Cannon Bailey, *Seaman A. Knapp, Schoolmaster of American Agriculture* (New York: Columbia University Press, 1945), 99, 101. Bessey explained the Hatch Act and its application to Nebraska in Nebraska State Horticultural Society Annual Report (1888): 158–162.

2. "An Act to Establish Agricultural Experiment Stations [Hatch Act]," in *The Statutes at Large of the United States of America* (1887; repr. Washington, D.C.: Government Printing Office, 1970), 24: 440–442.

3. Richard A. Overfield, *Science with Practice: Charles E. Bessey and the Maturing of American Botany* (Ames: Iowa State University Press, 1993), 3–13.

4. Quoted in ibid., 48.

5. Charles E. Bessey, "The Plan of Work in the Department of Horticulture in the Industrial College of the University of Nebraska," Nebraska State Horticultural Society Annual Report (1886): 108.

6. Ibid., 111; Overfield, *Science with Practice*, 48–49.

7. Bessey, "Industrial Education," Nebraska State Board of Agriculture Annual Report (1885): 81; second Bessey quote in Overfield, *Science with Practice*, 71.

8. Charles E. Bessey, *Botany for High Schools and Colleges*, 6th ed. (New York: Henry Holt, 1889); Bessey, *Elementary Botanical Exercises for Public Schools and Private Study* (Lincoln: J. H. Miller, 1894); Bessey, "High School Botany," *Science* 7 (1898): 266–267; Bessey, "Agriculture in the Common Schools," Nebraska State Board of Agriculture Annual Report (1899): 29–32 (quote on p. 32).

9. Charles E. Bessey in *Nebraska Farmer* 14, no. 5 (January 30, 1890): 89; Donald M. Cleary to Bessey, Beatrice, February 5, 1906, and Bessey to Cleary, Lincoln, February 13, 1906 (microfilm), reel 16, Bessey Papers, University of Nebraska, Lincoln [hereafter cited as Bessey Papers].

10. Charles E. Bessey, "Cryptograms or the Fungi Growth of Plants," Nebraska State Horticultural Society Annual Report (1885): 45; Bessey, "Report of the Botanist upon the Grasses and Forage Plants of Nebraska," Nebraska State Board of Agriculture Annual Report (1888): 131–142.

11. Overfield, *Science with Practice*, 54–57.

12. Ibid., 60–63.

13. Ibid., 65, 132–133; Charles E. Bessey, "Progress of the Botanical Survey of Nebraska," *American Naturalist* 29 (1895): 580–582.

14. Charles E. Bessey, "A Meeting-Place for Two Floras," *Bulletin of the Torrey Botanical Club* 14 (1887): 189.

15. Bessey, "A Preliminary Report upon the Native Trees and Shrubs of Nebraska," Nebraska Agricultural Experiment Station Bulletin (1892): 191.

16. G. A. Heywood, Gordon, Nebraska, to Charles E. Bessey [Fall 1891?], Bessey Papers, reel 4.

17. Charles E. Bessey and Albert F. Woods, "The Botany of the Apple Tree," Nebraska State Horticultural Society Annual Report (1894): 7–36; Bessey, "Botany of the Grape," Nebraska State Horticultural Society Annual Report (1895): 8–26;

Bessey, "Botany of the Plums and Cherries," Nebraska State Horticultural Society Annual Report (1895): 163–178; Bessey, "Notes on the Botany of the Strawberry," Nebraska Horticultural Society Annual Report (1896): 237–240.

18. Charles E. Bessey, "The Russian-thistle in Nebraska," Nebraska Agricultural Experiment Station Bulletin 31 (1893): 67–77; Bessey, "Report of the Botanist," Nebraska State Board of Agriculture Annual Report (1893): 75, 93–94 (quote on p. 94); Bessey, "Governor [Lorenzo] Crounse upon the Russian-thistle," Nebraska State Horticultural Society Annual Report (1894): 133–135.

19. Per Axel Rydberg, "Flora of the Sandhills of Nebraska," *Contributions from the U.S. National Herbarium* (USDA, Division of Botany), 3, no. 3 (September 14, 1895): 135. See also Paul A. Johnsgard, *This Fragile Land: A Natural History of the Nebraska Sandhills* (Lincoln: University of Nebraska Press, 1995).

20. Charles E. Bessey, unpublished manuscript, June 1, 1912, quoted in Raymond J. Poole, "Fifty Years on the Nebraska National Forest," *Nebraska History* 34 (September 1953): 144–145.

21. Bessey, "The Reforesting of the Sand Hills," Nebraska State Board of Agriculture Annual Report (1893): 95–96; Bessey, "A Preliminary Report upon the Native Trees and Shrubs of Nebraska," Nebraska Agricultural Extension Bulletin (1890): 171; J. C. Toliver to Bessey, Ainsworth, Nebraska, October 22, 1897, Bessey Papers, reel 8.

22. Charles E. Bessey, "A Suggestion as to State Forests," Nebraska State Board of Agriculture Annual Report (1892): 199–200; Bessey, "Reforesting of the Sand Hills," 97; Bessey, "The Forests and Trees of Nebraska," Nebraska State Board of Agriculture Annual Report (1899): 79–102; see also John Wesley Powell, *Report on the Lands of the Arid Region of the United States*, ed. Wallace Stegner (Cambridge: Harvard University Press, 1962), 26.

23. Poole, "Fifty Years on the Nebraska National Forest," 145–146.

24. William K. Hall, "The Investigation Now Being Made in Nebraska by the U.S. Bureau of Forestry," Nebraska State Horticultural Society Annual Report (1902): 149–159; Overfield, *Science with Practice*, 152–153.

25. Poole, "Fifty Years on the Nebraska National Forest, 155.

26. Ibid., 155–159; C. A. Scott, "The Work of the Bureau of Forestry in Nebraska," Nebraska State Horticultural Society Annual Report (1904): 11–15.

27. B. F. Gentry, "First Irrigation in North Platte River Valley," in Asa B. Wood, ed., *Pioneer Tales of the North Platte Valley and the Nebraska Panhandle* (Gering, Neb.: Courier, 1938), 214; Rydberg, "Flora of the Sand Hills of Nebraska," 144–145; see also W. P. Snyder, "Dry Land Farming in Western Nebraska," Nebraska State Board of Agriculture Annual Report (1909): 218.

28. Charles E. Bessey, "Some Agricultural Possibilities for Western Nebraska," Nebraska State Board of Agriculture Annual Report (1900): 76, 109; J. Sterling Morton to Bessey, Nebraska City, January 6, 1902, Bessey Papers, reel 11.

29. *Statutes* 33: 547; Everett Dick, *Conquering the Great American Desert: Nebraska* (Lincoln: Nebraska State Historical Society, 1975), 135.

30. Charles E. Bessey, "Address of Welcome," Nebraska State Horticultural Society Annual Report (1907): 78.

31. "Organized Agriculture Banquet in Honor of Dean Charles E. Bessey, University of Nebraska," Nebraska State Board of Agriculture Annual Report (1913): 158.

32. Ibid., 164–168, 175.

6

Creating Home on the Range

Wyoming is like no place else on earth. On the one hand, it holds a fateful attraction that creates an almost mystical commitment; on the other hand, it engenders a tedious annoyance that creates feebleness and tempts departure. Adhering to the former and avoiding the latter has been the key to thriving in Wyoming. And while Aven Nelson failed to turn Wyoming into a horticultural paradise, he left a remarkable legacy, not the least of which was to demonstrate the state's horticultural possibilities and to argue for their usefulness in making Wyoming a more livable place.

Nelson wanted Wyoming to grow from a vast expanse inhabited by transient cowboys to a place of permanent homemakers. In 1911, by which time his reputation as a friend of rural residents was well established, Nelson read a paper titled "Making a Home, or Staying on the Place," to the Young Men's Literary Club of Cheyenne, whose members included the city's leading citizens, many of them ranch owners. He began, as he often did, by invoking

Scripture, in which Ruth (Ruth 1:16–17) declared her willingness to emigrate to a new home, as well as to die and be buried there. He contrasted that with Wyoming: "[W]e have been unwilling to die here, or in case the summons came unexpectedly we were even then unwilling to be buried in its soil."

More directly to his horticultural point, Nelson continued, "[F]orty years [since Wyoming was separated from Dakota Territory] have not served to wipe out the pictures made by pen or word that depicted Wyoming as the place where the Creator dumped the refuse when he had finished fashioning the world and called it good. For thirty years the press of this nation had printed and reprinted Bill Nye's funny but fatal fallacies. Outside of Wyoming the press will recall his terse description of our seasons, 'ten months winter and two months late fall.'" Nelson also noted Nye's famous etching of Wyoming's lone tree, after which—again in contrast—he displayed photographs of the Wyoming orchards he had inspected. He asked club members not simply to help "set Wyoming right before the world" but to prove Bill Nye wrong by growing more within their own community.[1]

Nelson's appeal for community horticulture, with emphasis on its economic benefits, was picked up and spread across the High Plains by *Field and Farm*, the region's leading agricultural newspaper: "The village, town or city that does not in its collective capacity, give any attention to the beautifying of its streets, or the development of a park, or park system, is falling short of its opportunities for usefulness. There is no one thing that will do more to advance the value of towns and city property in general than those efforts which tend to improve the appearance by making it uniformly attractive and inviting to the eye." Like Charles Bessey, his senior by fourteen years, Nelson preached "the gospel of a higher horticulture" rather than that which had to do with its commercial aspects alone.[2]

While Nelson would be recognized as the preeminent botanist of the western United States and earned an international reputation for his scientific research, it is his application of botany to the horticulture of the High Plains that concerns us here. Like Bessey, Nelson viewed himself as a teacher first and foremost; and, in the finest land-grant college tradition, he too meant for his teaching to reach far beyond the confines of the formal classroom. Toward that end, he contributed dozens of horticultural articles to a variety of Wyoming publications, traveled indefatigably to meet farmers and ranchers at their places, addressed innumerable agricultural and civic groups, and worked with a long succession of governors and legislators—making him, in every sense of the term, a missionary for horticulture.

Aven Nelson (1859–1952) was born on a farm in southeast Iowa. According to his biographer, he gained from his father a reverence for life and an instinct for beauty. He taught primary school before earning a B.A. from Missouri State Normal College (now Truman State University) at Kirksville, taking his first college teaching position in English at Drury College in Springfield, and being appointed superintendent of schools at Ferguson near St. Louis. On Arbor Day 1887, Nelson displeased Ferguson's school trustees and some parents by allowing all students, including girls, to plant trees with their own hands on the school grounds. By then, he had applied to Wyoming University and was hired to teach English soon after the Ferguson trustees declined to renew his contract.

When Nelson, age twenty-eight, arrived in Laramie in the summer of 1887—the first faculty member to have formally accepted an offer from the university—he discovered that the board had inadvertently hired two English professors out of a total staff of six. Since the other English professor had a master's and Nelson only a bachelor's degree, Nelson agreed to be the biology professor. His preparation for this assignment consisted of a lifelong interest in natural history, his annual springtime search for flowers, six lectures on plants he had heard at Missouri State Normal College, and occasionally assisting the biology professor at Drury.

To qualify as a land-grant institution, Wyoming had to list a variety of courses, which meant Nelson's initial academic responsibility, besides biology, included teaching physical geography in the School of Mines and economic botany, zoology, animal physiology, and hygiene in the Agricultural Department, as well as serving as university librarian and instructor of calisthenics. With the establishment of the Wyoming Agricultural Experiment Station during the university's fourth academic year (1890–1891), Nelson was appointed station botanist, and his professorial title expanded to include botany and horticulture.[3]

To assist Nelson, the university's board hired Burt C. Buffum, a recent graduate of Colorado Agricultural College, as instructor in agriculture and horticulture with the additional responsibility of university groundskeeper, the campus still indistinguishable from the rest of the Laramie Plains. Buffum also took over Nelson's courses while Nelson spent time at Harvard, earning an M.A. in botany in 1892.

Nelson and Buffum worked well together, although their approaches to horticulture differed greatly. Nelson, like Bessey, premised his approach on the conviction that there is only one form of science and that horticulture involves the application of its principles. Buffum, however, seemed more comfortable with the notion that there are two forms of science—pure and applied—and that horticulture as an applied science could be taught and practiced pretty much independent of an understanding of nature's laws.

While Nebraska, as a result of Bessey's efforts, had managed to forestall such a two-tiered approach, Wyoming introduced "applied science" virtually from its beginning. Buffum taught the agricultural course, which included "the history, cultivation, and propagation of garden vegetables" and "landscape gardening." In 1895–1896, the university required all freshmen and sophomores to enroll in "practical farming" and offered a one-year course of study "for those whose time is limited, but who wish some practical instruction in farming and ranching."[4]

Given the attractiveness of the present-day Wyoming campus, it is hard to imagine that it began as a treeless, sagebrush meadow. In his working plan for landscaping, dated May 1891, Buffum reported that he had on hand for trial several species of trees and shrubs from local residents, Colorado nurserymen, and Colorado Agricultural College. Finding a variety of trees for different landscaping purposes was far less significant than finding those that would survive the rigors of the local climate. Thus, the native cottonwood (*Populus deltoides* Marsh.), "despised in some localities on account of its hardy character," took first place for "all but lawn decoration," followed by other poplar species: narrowleaf cottonwood (*Populus angustifolia* James), balm-of-gilead (*Populus jackii* Sarg.), and Carolina poplar (*Populus x canadensis* Moench).[5]

For trials in horticulture, the campus with its trees and shrubs complemented the college farm where Buffum began by planting several varieties of garden vegetables including peas, beans, onions, and potatoes. In addition, Nelson suggested that it might be worthwhile to try cultivated varieties of eight small native fruits he had found along the hills east of Laramie, which he identified simply as strawberry, red raspberry, gooseberry, currant, buffaloberry, barberry, serviceberry, and wild cherry.[6]

In part because of the variations in elevation throughout Wyoming, ranging from 7,200 feet at Laramie to 3,750 feet at Sheridan, the university's board interpreted the Hatch Act as allowing for the establishment of experimental farms or substations throughout the state as part of the single Wyoming Agricultural Experiment Station. For the High Plains region of southeastern Wyoming, the university received forty acres near Wheatland, under Ditch No. 2 along the Cheyenne and Northern Railway line, from the Wyoming Development Company. This private venture had been organized in 1883 by Joseph M. Carey to divert waters of the Laramie River for irrigating more than 50,000 acres around Wheatland. This was the same Carey who later became

U.S. senator, governor, and Aven Nelson's most influential supporter in promoting horticulture in Wyoming.

The Wheatland Experiment Farm—at an elevation of 4,750 feet, or 2,500 feet lower than Laramie and on the eastern slope of the foothills—experienced a longer growing season, somewhat more precipitation, and, of course, reliable irrigation. Under Buffum's supervision, workers strung a barbed wire fence around the farm, constructed a barn and a toolshed, and purchased horses, equipment, and tools. Sod was broken to a depth of five inches, thoroughly cultivated, and then irrigated to rot the grass roots. Of the thirty-six square-acre plots into which the farm was divided, three were set apart for orchard and small fruits, two for vegetables and flax (for linseed oil), and one for shade trees and shrubs; the rest of the plots were reserved for grain and forage crops.

Planting got a late start the first year (1891), with unknown varieties of peas, corn, potatoes, turnips, watermelon, muskmelon, pumpkin, apples, cherries, and plums, as well as willows and cottonwoods for shade trees. First-year harvests and yields, if any, were not recorded. During the second year, beets, turnips, rutabagas, carrots, and potatoes did exceedingly well. In the third year, heavy spring winds followed by a hot, dry summer generally resulted in very poor results, despite an unusual amount of irrigation. The only exception was potatoes; to the popular question of whether it was better to use whole or cut specimens as seed potatoes, Wheatland demonstrated in favor of whole potatoes. Several unspecified varieties of plums and cherries bore fruit within three years of their planting; unspecified varieties of currants, gooseberries, and raspberries fruited fairly well. Favorable conditions in late March of the sixth season (1896) heralded a promising outlook until a shortage of irrigation water in late April and early May, combined with a hailstorm, extensively damaged all plants and destroyed all fruit blooms. Again, potatoes managed to produce a fair yield.[7]

Despite these setbacks, Martin R. Johnston, the farm superintendent, reported receiving more and more visitors from throughout the region, responding to requests for specific information, and generally concluding that the Wheatland farm was having "a very stimulating effect on [the High Plains of] eastern Wyoming." Indeed, following extension of irrigation by the Wyoming Development Company (under the Carey Act of 1894), James F. Jenkins of Cheyenne used the publication *Herald and Presbyter* to float the idea of establishing a church-related agricultural colony at Wheatland. While this did not occur, a Denver-area land developer who read about the idea did locate ten Greeley families on the Wheatland flats. Each family purchased a

160-acre plot; they all understood how to farm under irrigation and welcomed lower land prices than those around Greeley.[8]

Buffum, meanwhile, had pursued investigations into the possibilities of growing fruit trees and small fruits, whether irrigated or not, in eastern Wyoming. He began with a questionnaire mailed to farmers and ranchers in 1892. From the twenty-one responses, he concluded that fruit growing in eastern Wyoming was virtually nonexistent. Two notable exceptions were M. J. Goodwin of Lusk, who was growing several varieties of apples and small fruit, and John H. Gordon of lower Horse Creek near Cheyenne. "I am fully persuaded," Gordon wrote Buffum, "that all fruits which can be grown in Northern Colorado can be grown in this section, provided the same care and attention is given." Such care included late-fall irrigation to lessen drying out during the winter; mulching trees, taking care that the mulch did not touch the tree trunks; and gently turning down the branches of small fruits and covering them with soil, again for winter protection. Based on five years of demonstration work at Wheatland, Buffum concluded that both orchard and small fruits could do well throughout the High Plains.[9]

Like many substations across the country, Wheatland served as a demonstration farm rather than a true experiment station. In 1897 the new secretary of agriculture, James "Tama Jim" Wilson, former director of the experiment station and professor of agriculture at Iowa Agricultural College, prohibited the use of Hatch funds for support of such substations. As a result of his action, combined with the unwillingness of the Wyoming Legislature to pick up the funding, work at the Wheatland farm was discontinued after eight years, and the property returned to the Wyoming Development Company.[10]

Despite its short existence (1891–1898), the Wheatland farm did demonstrate that horticulture could succeed on Wyoming's high plains. As horticulture developed, however, Wyoming began to experience some of the "evils" long confronted by cultivators farther east. It was no surprise, therefore, that Aven Nelson's earliest forays into economic botany, like those of Charles Bessey, concerned plant diseases and imported weeds. Nelson, too, wrote with remarkable clarity for those whose reading had not been along botanical lines, briefly explaining general principles and causes, then describing characteristics to help with identification, and, finally, suggesting preventive measures.

For human nourishment, according to Nelson, no plant was more useful, or more successfully grown with a minimum of expense and labor, than the potato. Its most serious disease is scab (*Oospora scabies* Thaxter., now known as *Streptomyces* spp.), which causes surface blemishes and, thereby, severe economic damage. Moreover, once scab is introduced into the soil, it persists for several years, even after potatoes are no longer planted. Because scab was thought to be introduced exclusively through seed potatoes, Nelson argued that the first preventive step was to keep diseased potatoes out of Wyoming, although he did not specify how that could be accomplished. He did, however, outline a treatment to ensure that, before being planted, seed potatoes would be scab-free: "In about fifteen gallons of water dissolve two ounces of corrosive sublimate (bichloride of mercury). In this solution immerse the seed potatoes for one and a half hours, after which spread them out to dry, then cut and plant as usual." While seed potato disinfectants are no longer used, soils are sometimes treated before planting. By reducing alkalinity and increasing acidity, and by maintaining proper moisture during early tuber development, scab can generally be adequately controlled.[11]

Of greater threat to Wyoming horticulture was the Russian-thistle. In 1894, the year after Charles Bessey warned Nebraskans about its rapid spread, Aven Nelson had positively identified Russian-thistle near Cheyenne. Consequently, Nelson wrote the Wyoming Experiment Station's first "press bulletin," a short article meant for republication in local newspapers, on the topic of the Russian-thistle. Like Bessey, he noted that it was not a thistle, so called only on account of its spines; he added that it closely resembled the common tumbleweed (*Amaranthus albus* L.) and grew especially well in Wyoming's dry alkaline soils. While there was still time to forestall a full-scale invasion, Nelson urged immediate preventive measures: "If farmers and ranchmen and road commissioners will look out for it and destroy all suspected plants by cutting if green, by burning if mature, and if all railway lines will see that the same is done on their right-of-way, as I learn the Union Pacific is doing, we shall be spared the trouble, the injury and the loss that our sister States are undergoing." Nelson also suggested that the time was opportune for the state legislature to enact a weed law such as the one adopted in Wisconsin. Meanwhile, he solicited correspondence on the subject from residents, invited them to mail him suspected plants for determination, and added that "other noxious weeds will also cheerfully be determined."[12]

Nelson identified the Russian-thistle as the most dangerous weed in Wyoming, a plant with no redeeming features, able to choke out native grasses and take over cultivated lands. Like so many of his horticultural writings, his 1896 disquisition against around fifty different species of Wyoming weeds took on a distinctly moral tone. "The weediness of a plant," he wrote, "depends in many cases upon the observer's point of view; that is, a plant is pronounced a weed, bad in proportion to the hindrance it offers to the production of that which is at that particular time and place the desirable, or good." And weeds, he continued, are like the poor, always with us, "and again like the poor, they are the most numerous in the most shiftless communities," by which he meant in the most disturbed soils.[13]

The Wyoming Legislature had passed a law, signed by Governor William A. Richards on February 26, 1895, against the Russian-thistle, but the legislature provided neither the means to publicize its danger nor information on how to recognize it. Nelson unsuccessfully urged lawmakers to appropriate a few hundred dollars to support prevention efforts through a newly formed weed commission. His moralistic statement on weediness, however, may have worked against him because, under pressure from stock growers, the legislature repealed the anti-thistle law in 1903. Wyoming stock growers argued that properly cut and cured, Russian-thistle made fine fodder. Indeed, around the turn of the century, stockmen in eastern Colorado, which had no weed law, were mixing Russian-thistle and native hay. From their viewpoint, when choices were limited, a stack of thistles beat a snowbank in providing winter forage for livestock.[14]

Although the Wyoming Agricultural Experiment Station devoted more and more of its resources to the livestock industry, Aven Nelson limited his public service to the application of botany to horticulture. Very early, he had observed that, with sufficient irrigation, trees and shrubs newly planted at higher altitudes on the High Plains did well during their first growing season but often did not survive their first winter. In view of the fact that New England and Upper Midwest winters were as cold as those on the High Plains but winter killing was far less common, Nelson looked for causes of winter kill other than low temperatures. During his stay at Harvard, Nelson prepared his second experiment station publication, "The Winter-Killing of Trees and Shrubs."

Whether it was because it was early in Nelson's botanical career, because he had started as an English teacher, or because of his strong commitment to "useful knowledge"—or, most likely, for all these reasons—his brief description of the functions of the root, stem, and leaf serves as an eminently intelligible introduction to the main facts of plant nutrition. Again in lay terms, he explained the phenomena of desiccation and transpiration and suggested that the remedy to winter kill, easily seen but often difficult to apply, was to keep plants, especially their roots, moist. His suggestions, with explanation of both their usefulness and applications, included late-fall watering and, if possible, some watering during winter months; mulching, regardless of whether irrigating was possible; and wrapping or otherwise protecting stems and branches of tender or somewhat hardy trees and shrubs during their first winter or two.[15]

As a result of his own fieldwork on Wyoming plants, Nelson prepared an experiment station bulletin "The Trees of Wyoming," meant not simply to teach his fellow citizens how to identify trees but also to cultivate an interest in nature and, in particular, its civilizing aspects for home and community. While most of us are familiar with its sterner aspects—the bleak wind-swept plain, the roaring blizzard—we are nearly all strangers to its lovelier traits. For evidence of this apparent indifference, Nelson pointed to towns "without a tree or a shrub or even a bit of lawn." As origin of this condition, he suggested that early settlers had been too busy to plant and viewed their stay in Wyoming as temporary. But Wyoming was becoming less of a camping ground, so the time was ripe for Wyomingites to join the great wave of tree-planting enthusiasm sweeping the country. He referred in particular to Nebraska, where "transformation in one generation from dreary treelessness to restful though animated homelikeness must be considered one of the achievements of the age."[16]

Nelson attributed the awakening of public interest in tree planting to the schools, where children and, by extension, their parents were being taught about nature, including how to identify the plants around them. Ever the teacher, Nelson anticipated the novice's frustration with nomenclature. How desirable it would be to have a good, expressive English name, a single common name on which we all could unite, not only for our trees but for all our plants. The difficulty, however, is twofold: most plants have different common names in different localities, and, what is worse, very different plants have the same common name. To illustrate, the lodgepole pine is also called tamarack and black pine in Wyoming, spruce pine in Colorado, and by other names in other places. To complicate matters further, these names are also applied to other trees, as, for example, tamarack to another species of pine and to two different species of larches. How, then, Nelson asks, can we know what tree is meant when we hear a name?

The answer is to use its Latin or scientific name because, the world over, only one such name can be applied to the plant in question. Such a name consists of two parts—the first, or generic, name, which in a way, he explained, corresponds to a person's family name; and the second, or specific, name, "which points out a particular kind much as a given name points out a particular person." To illustrate, the pines belong to the genus *Pinus*; to indicate a particular kind or species of pine, a second word is added, such as *Pinus contorta* for the lodgepole pine. It is also customary to add the name, or its abbreviation, of the person who first described the species or brought the correct generic and specific names together; thus, lodgepole pine would be *Pinus contorta* Dougl.

To those who would think Latin names are difficult to learn or to understand, Nelson suggested that "they are in fact more easily remembered than common names for they are often descriptive of some character of the plant."[17] And, with regard to native trees, another encouraging note is that the small number of genera and species makes for the least intimidating introduction to the use of plant classification keys.

As with Bessey before him, Nelson's horticultural contributions were founded on his botanical research. Far more than Bessey, however, Nelson himself collected the plants he studied and, in the course of his career, proposed hundreds of new genera, species, and varieties (subspecies). To help classify plants and prepare for exchanges with other institutions, Nelson was able to hire Elias Nelson (no relationship), who was majoring in botany, as field and herbarium assistant, beginning in 1896. The specimens they started collecting in the 1890s, dried, and systematically arranged for reference were the beginning of what is now recognized as the preeminent repository of Rocky Mountain plants, second only to the St. Petersburg herbarium in its collection of alpine plants.

By 1899, Elias Nelson had earned the first master's degree ever awarded by the University of Wyoming—having written his thesis on the genus Phlox—and taken the position of assistant horticulturist in the experiment station, assuming some duties formerly held by Burt Buffum, who had resigned to take a faculty position in agriculture at Colorado Agricultural College. Meanwhile, at the new Wyoming Experiment Station greenhouse, Elias Nelson began a prodigious program of trials on garden vegetables, flowering ornamentals, small fruits, shrubs, and trees; he also prepared Wyoming's horticultural exhibit for the Louisiana Purchase Centennial Exposition in St. Louis. In his publication on the shrubs of Wyoming, he emphasized their value for preservation of watercourses, as stock shelter, and as windbreaks and expressed the hope that native fruits would soon be productively cultivated.[18]

In 1902, Aven Nelson published an experiment station bulletin on native vines, which, in the words of his biographer, "has lost none of its pertinence with the passage of time." Indeed, no better argument has been made on the eminent utility of beautifying home and community. Although written for Wyoming readers, it pertained to the entire region. The editor of *Field and Farm* requested some of Nelson's photographic plates for reproduction to show readers what could be done to beautify homes on the High Plains.[19]

Nelson began his bulletin on native vines with a discourse on a citizenry that, in his day, still considered itself largely temporary, exploiting Wyoming's resources and, after amassing whatever sum seemed to constitute wealth, returning to the older states. Outside of towns there were few homes but many temporary camps; in towns, frame houses were often unpainted, unfenced, dilapidated, and—referring to the common 25- by 115-foot town lots—crowded together despite Wyoming's wide-open spaces: "No green lawn to relieve the eye from the glare reflected from the heated hills; no tree to break the monotony of the interminable plains . . . no bed of flowers to gratify the aesthetic sense present in some degree even in the worst of us."[20]

Ever the optimist, Nelson did see a trend gradually but surely changing Wyoming from a vast camping ground into a commonwealth of pleasant homes. For one thing, there was a new generation, born in Wyoming, with no ties to other states. Second, he saw fewer speculators and more long-term investors, seeking modest returns for well-directed efforts in the trades, farming, and stock raising.

Regarding the prevailing attitude that any attempt to grow ornamentals represented time, money, and labor wasted, Nelson noted that the most common cause of failures in the past, of which there were many, was the use of plants not acclimatized to local conditions. We should learn from these failures that while Wyoming is no colder than many other states, its overall climate is different: consider its air density and moisture, air currents, and intensity of light and day/night temperature inequalities. Considering that plants adapt to new conditions over many generations, much more slowly than animals do, we should not expect plants from a wholly different environment to survive, especially after prolonged exposure during transportation and often indifferent planting. To put it starkly, an Inuit transplanted from Alaska to Guatemala would have a better chance to adapt than a rose transplanted from Alabama to Wyoming.

Successful planting, accordingly, depends on obtaining the right variety of seed or plant, which in most cases means northern-grown—the nearer to Wyoming growing conditions, the better. Indeed, Nelson encouraged the use of native plants or nursery-grown plants from native stock. While Wyoming has only a few native vine varieties and none unique to Wyoming, no group of plants gives itself more readily to training or serves a better purpose in making the home attractive. Found in all the High Plains states, the common vines are Virginia creeper (*Parthenocissus quinquefolia* [L.] Planch.), western clematis (*Clematis ligusticifolia* Nutt.), river-bank grape (*Vitis riparia* Michx.), common hops (*Humulus lupulus* L.), and wild cucumber (*Echinocystis lobata* (Michx.) T. & G.).

Lest anyone think the impact of pleasant surroundings is limited simply to appeasing our aesthetic nature, Nelson argued that attractiveness positively influences each person's productive capacity and affects that person's relationship to the community as a whole: "When the individual and the municipality shall heartily join hands for the beautifying of the home and of the city a great stride will have been taken towards better citizenship."[21]

Nelson's article on vines elicited favorable responses not only from the agricultural press but also from individual growers. John H. Gordon of lower Horse Creek particularly enjoyed the article as he took great pleasure in gazing on his "lovely Creepers," even though they were hugging his trees to death. A former Greeley potato farmer turned stock farmer, Gordon had corresponded earlier with Buffum on the subject of fruit trees; in 1909, he became the irrigation farmer-in-charge at the United States Department of Agriculture experimental farm near Cheyenne.[22]

Nelson's tireless work to promote horticulture in Wyoming took place in addition to his principal occupation as professor of botany. During the first decade of the twentieth century, he was engaged in revising John Merle Coulter's pioneering work, *Manual of the Botany of the Rocky Mountains*. Although he was gaining a national reputation, his ongoing frustration with a wholly unenlightened university administration led him to send, unbeknownst to anyone in Wyoming, a stunningly candid five-page handwritten letter to Charles Bessey. Nelson had belatedly learned about a resignation in botany at Nebraska; Bessey had the unpleasant task of responding that the position had been filled. Unquestionably, Nebraska would have hired Nelson had his wish to leave Wyoming been known earlier.[23] Such wishes, incidentally, continue to affect from time to time anyone who seeks to make Wyoming better but who is not always prepared to cope with the entrenched opposition to change.

While we can only imagine the nature of opposition to Nelson's civic efforts to promote horticulture, he was able to proclaim that the Wyoming Legislature had accomplished "a very rare [feat] in which the stable was locked before the horse was stolen." By an act of February 15, 1905, the legislature had created a State Board of Horticulture with regulatory authority to protect Wyoming orchards. While Nelson welcomed the legislation as a sign that Wyoming was no longer dominated by the "great cattle outfits" and that the state welcomed the advent of farms, his optimism turned out to be premature.[24]

The Wyoming Horticulture Act provided for inspection and, where necessary, disinfection or destruction of nursery stock entering the state. Specifically, out-of-state nurseries wishing to do business in Wyoming had to file an affidavit with the board affirming that their stock had been inspected by a competent inspector and pay a $25 license fee, good for two years, after which they received Wyoming shipping tags identifying both place of origin and destination of their shipments. In addition, each out-of-state nursery had to post a $500 bond.

At its organizational meeting on December 6, 1905, the six-member Board of Horticulture elected Governor B. B. Brooks as permanent chair and Nelson as permanent secretary with the title "Chief Inspector of Fruit Pests." Each of the remaining four members served as "Inspector of Fruit Pests" in his respective district of the state. The legislature initially allocated $1,000 per year to the board, out of which Nelson received $50 per month for conducting all staff work. Presumably, the remaining $400 for supplies, travel, and other expenses would be supplemented by revenues from the sale of licenses.

Nineteen out-of-state and the two known in-state nurseries received licenses during the board's first year of operation. Several others declined to pay the regulatory costs—some on the grounds that the small size of their business in Wyoming made those costs prohibitive, and others on the specious grounds that the Wyoming law was contrary to the U.S. Constitution's provision against commercial discrimination between the states.

Nelson, meanwhile, had written the board regulations, based on information he had gathered from other states and Canadian provinces. He included an appeal for public cooperation with his pledge of assistance: "Fruit growers and owners of even the smallest orchards are earnestly requested to report to the Secretary of the Board the condition of their orchards. If insects or fungous diseases have found entrance, it will be the duty of the Secretary to co-operate with the owner for the extermination or eradication of such insects or diseases. To that end, printed directions for the treatment of infested or infected trees will be sent and all other available information will be supplied."[25]

The number of Nelson's popular publications that encouraged fruit growing and horticulture in general was amazing. It included the Horticulture Board's biennial reports to the legislature, six board-sponsored special bulletins, numerous newspaper articles, and his regular horticultural column in the *Ranchman's Reminder*—a monthly publication of the Wyoming Experiment Station—and its successor, *The Wyoming Farm Bulletin*. During 1905–1906, Nelson personally mailed out 1,000 copies of the board's enabling legislation and regulations to every Wyoming newspaper, county clerk (multiple copies for posting in public places), county commissioners, other public officials, and interested citizens as well as the principal railroad, express, and stage offices in the state.

Orchard, Wheatland, Wyoming, 1913. Courtesy, Aven Nelson Papers, American Heritage Center, University of Wyoming, Laramie.

As Board of Horticulture secretary, Nelson personally handled all correspondence concerning licensure as well as a much larger, seemingly overwhelming correspondence seeking general information. He welcomed the latter, first, because his office was state supported for the express purpose of supplying horticultural information and, second, because it indicated to him a steady growth of interest toward permanent residency. In addition to office work, Nelson spent time in the field, not only on inspection trips to orchards but also to participate in Farmers' Institutes, the state fair, and other agricultural-related events. At least one special trip was made to Converse County, in his words, "to straighten out a difficulty in regard to some shipments concerning which there was some misunderstanding." So far as is known, the board voted to revoke only one nursery license, in 1906, following incontrovertible evidence that the unnamed nursery had knowingly violated Wyoming law by shipping into the state a consignment of apple trees infected with wooly aphids.[26]

In his biennial reports, Nelson was able to present the legislature with county-by-county summaries of the progress of horticulture. In his first report, for example, he noted that Laramie County, which at the time included most of the southeastern Wyoming plains, had provided outstanding fruit entries

Niobrara County produce, state fair, Douglas, Wyoming, 1913. Courtesy, Aven Nelson Papers, American Heritage Center, University of Wyoming, Laramie.

at the second state fair. He also commended Wheatland, recognized as the most successful farming community on the High Plains of Wyoming, for turning its attention to fruit growing; in later reports he made reference to "dry-farming" experiments with fruit growing, both in eastern Wyoming and western Nebraska. From farmers in Laramie and Niobrara counties, he included details about establishing windbreaks, with minimal irrigation, before planting orchards.[27]

Regarding ornamentals, Nelson lavished praise on the residents of Cheyenne for supporting the finest city park in Wyoming and for their beautiful lawns and well-shaded grounds, often including fruit-bearing trees and shrubs. Horticulture, he reminded his readers, is not concerned merely with growing fruits but also with growing plants for ornamental purposes. To which he repeated: the community that gives no attention, in its collective capacity, to the beautifying of its streets or the development of parks is falling short of its "opportunities for usefulness."[28]

In setting out the duties of the State Board of Horticulture, legislators had included this instruction: "To call together and hold in conjunction with Horticultural Societies, public meetings of those interested in horticulture and kindred pursuits." For carrying out this provision, the legislature appropriated

Wealthy variety apple. Courtesy, Aven Nelson Papers, American Heritage Center, University of Wyoming, Laramie.

$200 per year, suggesting that horticulture may not have been uppermost on its agenda. Nonetheless, within the spirit of that provision, Nelson, in his capacity as secretary of the State Board of Horticulture, called an organizational meeting of the Wyoming State Horticultural Society, to be held during the 1907 state fair in Douglas. Although we have no figures, we can safely assume that the number in attendance was small. And, as so often happens with voluntary associations, Nelson as the driving force in getting the group started was enlisted to serve as its secretary-treasurer.[29]

Nelson envisioned the State Horticultural Society as the principal public vehicle through which to advance the interests of the garden, field, farm, and ranch. As if he were taking a page out of Tocqueville's *Democracy in America*, he reflected that this country "is much given to societies, associations and organizations for the purpose of promoting whatever may be the idea uppermost in the public mind at any given time." Nelson expressed with enthusiasm the belief that, by 1907, horticulture had reached that level among the citizens of Wyoming.[30]

Nelson, however, would be sorely disappointed because the Wyoming Horticultural Society never really caught on. Nelson himself had to convene the fourth annual meeting, in the unexplained absence of the president. Those present at the meeting were an unspecified "few," including a handful of visitors. Moreover, Nelson had sent a note to every member requesting—indeed, urging—a letter, paper, or address, but not a single paper or address was submitted. "This discouraging and practically hopeless condition," Nelson reported later, "came near [to] terminating the existence of the society."[31]

Luckily, Nelson had a letter from a loyal but absent member, containing several suggestions including the establishment of a journal through which members could communicate with each other. Because the society did not have the means to support such a publication, members welcomed the offer by Henry G. Knight, experiment station director, to make available, at no cost to the society, two or more pages per month in the station's *Wyoming Farm Bulletin* and to place every society member on its mailing list. When it came to designating someone to edit the articles, letters, notes, suggestions, and questions, the membership again turned to Nelson. He accepted very reluctantly and not before warning that if nothing were submitted there would be nothing for him to edit, and he would ask the station director to reassign the allotted print space.[32]

In December 1912, Director Knight gave over an entire issue of the *Wyoming Farm Bulletin* to horticulture, fruit growing in particular. His stated purpose was to provide the evidence "to convince anyone, who is open to conviction, that Bill Nye must have had another guess coming when he said the only thing one could raise in Wyoming was 'hell and heifers.'"[33] This did not stimulate new interest in the State Horticultural Society, however.

Nelson reckoned that part of the difficulty was confusion in the public mind regarding the respective roles of the State Board of Horticulture and the State Horticultural Society. In 1912, he explained that while the interest of both organizations was to encourage the fruit industry, the former was created by the legislature to protect the state against the introduction of injurious insects and serious fungus pests, and the latter was a voluntary association of citizens interested in growing "everything that tends to make their several communities more beautiful or profitable and therefore more desirable as places in which to live." The fact that the secretary of the Board of Horticulture also served as secretary-treasurer of the society, he added, was merely a coincidence resulting from his "desire to help all organizations that have for their object the making of a more beautiful and a better Wyoming."[34]

The failure to recruit enough members to make the society viable rendered Nelson all the more indispensable and helped ensure the society's demise when Nelson had to turn his attention to the job of acting president of the university in the summer of 1917. At the time, Frank Julian, a Casper-area rancher, society president, and strong Nelson supporter, wrote Nelson about membership: "I do not know just what we may do or what we may not do. But will unburden my mind to you." He sought Nelson's opinion on making the state's newspaper editors honorary society members and offering prizes at the state fair—for example, a first prize of two dollars for the largest apple and a second prize consisting of a year's membership in the society.[35]

But it was too late. Nelson's energies had been deflected to other matters. In his last known letter regarding the society, Nelson, ever positive, congratulated Julian on having secured attractive stationary for the society, and, as society treasurer, he offered to reimburse Julian upon receipt of a bill noted as "paid." Because he had received very little society correspondence during 1917, Nelson volunteered that he could use his State Board of Horticulture stationery. As to the suggestion of prizes, Nelson doubted that would get the society anywhere, but he ventured that there was no harm in trying, although he himself would not be able to pursue the matter with the state fair folks.[36]

If the Wyoming State Horticultural Society came to an end because of a lack of public interest, the Wyoming State Board of Horticulture dissolved largely because of the active opposition of the nursery industry and, we might suggest, the tacit opposition of a legislature dominated by members who favored priority use of water for forage crops essential to livestock.

Virtually from the beginning, as noted, some out-of-state nurserymen had objected to the board's fee structure either on the grounds that it made business in Wyoming uneconomical or that it unconstitutionally restrained the conduct of interstate commerce. The latter came to a head in 1913, when a Chicago law firm representing unspecified seed and nursery stock houses asked the District Court in Cheyenne (*E. S. Welch v. State Board of Horticulture*) to enjoin the board from carrying out the horticultural inspection law on the grounds that it violated Article 1, Section 10.2 ("Powers Prohibited to the States") of the U.S. Constitution. When the lower court upheld the board's authority, the plaintiff appealed to the State Supreme Court, which, on November 27, 1915, decided that the license fee was appropriate but that the bond was unconstitutional and thus could not be required. After which Wyoming Attorney General Douglas A. Preston, who had represented the board before the Supreme Court and was no admirer of Nelson, suggested that the board revise its regulations as soon as possible; he later reiterated his opinion that the board had exceeded its legislatively mandated authority by imposing unreasonable rules.[37]

Unquestionably, Nelson disagreed with the attorney general's opinion. Nonetheless, the legislature stripped the board of its authority in early 1917 (SF 58), with some of that authority entrusted to a new five-member Agricultural Advisory Board, for which the governor, the director of the experiment station, and the station botanist (Nelson) would serve as ex-officio or nonvoting members. In his final report to the Horticulture Board, Nelson, always positive in public, explained that the change had been made "in the interest of simplicity and economy." He did add the opinion, however, that as a result of the Horticulture Board's demise, horticultural work in Wyoming "quite probably" would be seriously handicapped.[38]

By 1919, the legislature had eliminated all funding in support of horticultural inspections by the Agricultural Advisory Board. Four years later, the legislature abolished the Board of Horticulture and the Agricultural Advisory Board and transferred to the University Board of Trustees all remaining powers, including that to repeal horticultural rules and regulations. At the same time, the legislature consolidated agencies assisting farmers and ranchers into a new Department of Agriculture. From that time on, agriculture in Wyoming meant raising livestock and related forage crops.[39]

While Aven Nelson had stated that his interest in, and commitment to, horticultural work would not diminish, the demise of both the State Board of Horticulture and the State Horticultural Society, combined with his taking on the university presidency, marked the end of his missionary work for horticulture. After the presidency, he returned full-time to the Botany Department, his professional interests intact and many botanical projects left to complete.[40] From then on, the university's role in the promotion of horticulture to homestead and community rested with a handful of committed county extension agents, formally recognized under the Smith-Lever Act of 1914. Among their precursors were college field workers experimenting with dry-land horticulture in eastern Colorado.

Notes

1. Aven Nelson, "Making a Home, or Staying on the Place," *Wyoming Tribune*, March 15, 1911, 3.

2. Secretary's report, *Wyoming State Board of Horticulture Special Bulletin* 3 (1912): 36–37, reprinted in *Field and Farm* 29, no. 1488 (August 8, 1914): 3; Bessey quoted in Nelson, *Wyoming Farm Bulletin* 2, no. 1 (July 1912): 187.

3. Roger L. Williams, *Aven Nelson of Wyoming* (Boulder: Colorado Associated University Press, 1984), 1–20.

4. Wyoming Agricultural Experiment Station Annual Report 1 (1891): 23, and 6 (1896): 13, 33.

5. B. C. Buffum, "Report of the Horticulturalist," *Wyoming Agricultural Experiment Station Bulletin* 1 (May 1891): 10.

6. Aven Nelson, "Report of the Botanist," *Wyoming Agricultural Experiment Station Bulletin* 1 (May 1891): 21.

7. Martin R. Johnston, "Report of the Wheatland Experiment Farm," Wyoming Agricultural Experiment Station Annual Report (1891): 37–39, (1892): 24, (1893): 19, (1895): 25–27, (1896): 27–28; James F. Jenkins, "My Life Story" [1925], copied by his daughter from the original, Jenkins Collection, ms. 7, 42, Wyoming State Archives, Laramie; William E. Smythe, *The Conquest of Arid America* (Seattle: University of Washington Press, 1969 [1899]), 227.

8. Johnston, "Report of the Wheatland Experiment Farm" (1892): 24; Jenkins, "My Life Story," 42.

9. Gordon quoted in Dice McClaren and B. C. Buffum, "Best Varieties and Breeds for Wyoming," *Wyoming Agricultural Experiment Station Bulletin* 5 (February 1892): 25; Buffum, "Fruit Growing in Wyoming," *Wyoming Agricultural Experiment Station Bulletin* 34 (1897): 144–145.

10. "Report of the Director," Wyoming Experiment Station Annual Report (1898): 53.

11. Aven Nelson, "The Grain Smuts and Potato Scab," *Wyoming Agricultural Experiment Station Bulletin* 21 (January 1895): 22–24 (quote on p. 24).

12. Aven Nelson, "The Russian-thistle," *Wyoming Agricultural Experiment Station Press Bulletin* 1, reprinted in Wyoming Agricultural Experiment Station Annual Report (1895): 29–31.

13. Aven Nelson, "The Worst Weeds of Wyoming and Suggested Weed Legislation," *Wyoming Agricultural Experiment Station Bulletin* 31 (December 1896): 267, 271.

14. Ibid., 311; *Field and Farm* 18, no. 897 (March 7, 1903): 8.

15. Aven Nelson, "The Winter-Killing of Trees and Shrubs," *Wyoming Agricultural Experiment Station Bulletin* 15 (December 1893): 213–219.

16. Aven Nelson, "The Trees of Wyoming and How to Know Them," *Wyoming Agricultural Experiment Station Bulletin* 40 (January 1899): 61, 64.

17. Ibid., 59–69.

18. Williams, *Aven Nelson*, 65–66, 77–78, 119; Wyoming Agricultural Experiment Station Annual Report (1903): 54–57; Elias E. Nelson, "The Shrubs of Wyoming," *Wyoming Agricultural Experiment Station Bulletin* 54 (July 1902): 1–4.

19. Williams, *Aven Nelson*, 113; *Field and Farm* editor to Wyoming Agricultural Experiment Station director, Denver, July 2, 1902, box 7, folder 2, Aven Nelson Papers, University of Wyoming, Laramie (hereafter cited as Nelson Papers).

20. Aven Nelson, "Native Vines in Wyoming Homes," *Wyoming Agricultural Experiment Station Bulletin* 50 (March 1902): 2.

21. Ibid., 1–15 (quote on p. 15).

22. John H. Gordon to Aven Nelson, Cheyenne, June 5, 1902, box 7, folder 2, Nelson Papers; John H. Gordon, "Experiments in Supplemental Irrigation with Small Water Supplies at Cheyenne, Wyoming, in 1909," USDA Office of Experiment Stations 95 (April 1910): 1–11.

23. Aven Nelson to Charles Bessey, Laramie, May 9, 1908, and Bessey to Nelson, Lincoln, May 23, 1908, reel 22, Bessey Papers, University of Nebraska, Lincoln.

24. "Report of the Botanist," Wyoming Agricultural Experiment Station Annual Report (1905): 21.

25. "Wyoming Horticultural Law," Wyoming State Board of Horticulture Biennial Report (1905–1906): 13.

26. Wyoming State Board of Horticulture First Biennial Report (1905–1906): 15, 27 (page of the quote), Second Biennial Report (1906–1907): 5–6, Third Biennial Report (1908–1909): 7, Fourth Biennial Report (1911–1912): 7.

27. Wyoming State Board of Horticulture, Third Biennial Report (1908–1909): 27, Fifth Biennial Report (1913–1914): 73–76.

28. Aven Nelson, "The Status of Horticulture in Wyoming," Wyoming State Board of Horticulture First Biennial Report (1905–1906): 48; Nelson, "Municipal Horticulture," *Wyoming State Board of Horticulture Special Bulletin* 3 (1912): 36–37.

29. *Session Laws of the State of Wyoming* (1905), Ch. 50, Sec. 5; Wyoming State Board of Horticulture Second Biennial Report (1907–1908): 16.

30. Wyoming State Board of Horticulture Fourth Biennial Report (1911–1912): 15.

31. "Horticultural Column," *Wyoming Farm Bulletin* 1, no. 5 (November 1911): 74.

32. Ibid., 74–75.

33. "Introduction," *Wyoming Farm Bulletin* 2, no. 6 (December 1912), 1.

34. "The State Horticultural Society," Wyoming State Board of Horticulture Fourth Biennial Report (1911–1912): 15–16.

35. Frank Julian to Aven Nelson, Casper, July 27, 1917, box 10, folder 3, Nelson Papers.

36. Frank Julian to Aven Nelson, Casper, January 17, 1918, and Aven Nelson to Frank Julian, Laramie, January 23, 1918, box 10, folder 3, Nelson Papers.

37. Douglas A. Preston, Cheyenne, to Aven Nelson, January 6 and 21, 1916, February 8, 1917, box 10, folder 3, Nelson Papers; Albert Marion Gaddy, "The Wyoming Department of Agriculture" (master's thesis, University of Wyoming, 1951), 9.

38. Secretary's Report, Wyoming State Board of Horticulture Seventh Biennial Report (1917–1918): 6–9.

39. Aven Nelson to Governor William B. Ross, Laramie, February 2, 1923, box 8, folder 10, Nelson Papers; *Wyoming Session Laws* (1923), Chs. 99, 100.

40. Williams, *Aven Nelson*, 218–219.

7

Limits of Dry-Land Horticulture

As a result of Secretary James Wilson's action to compel adherence to the research provision of the Hatch Act, Colorado Agricultural College closed its Cheyenne Wells substation, which had operated as a demonstration farm from 1894 through early 1900. The college reassigned James E. Payne, substation superintendent, to survey the fruits, vegetables, trees, and forage crops that grew on the homesteads of eastern Colorado. With horses and a spring wagon, Payne spent the summer of 1900 traveling over 1,000 miles, listening to homesteaders, inspecting their spreads, and keeping a detailed set of "Field Notes."

Payne's district lay within the Arkansas River valley on the south, the South Fork of the Republican River on the north (roughly parallel to Interstate 70), the town of Limon on the west, and the Kansas state line on the east. Today's casual traveler might conclude that this vast, rolling, treeless space now mostly covered with wheat and other cultivated grains is unfit for human habitation. Indeed, this area remains the least populated region on the High

Plains (fewer than two people per square mile), and precipitation is also the lowest, well under twelve inches per year. Combine that with the dramatic recent decline in the aquifer level and one might reasonably conclude that the Great American Desert does exist. But we are getting ahead of our story.

In introducing his 1900 "Field Notes," Payne noted that eastern Colorado had been rapidly populated during the wet years of 1868 and 1869. Settlers came mostly from Kansas and Nebraska, having been led to believe they could successfully cultivate using traditional techniques. Real estate "boomers" had enthusiastically referred to this area as the "Rainbelt" and argued that precipitation had been moving westward as grasslands gave way to cultivated fields. Between 1874 and 1894, however, cycles of catastrophic droughts, grasshopper plagues, severe winters, and financial panics caused at least three waves of migration out of this same country.[1]

By the mid-1890s, the few settlers who remained had learned that in order to survive, they needed to engage in a combination of approaches to conserve moisture: dig wells, build and maintain small reservoirs, and use special methods of tillage to preserve moisture in cultivated soil. The last approach became known as dry farming or dry-land farming.

Elwood Mead had plainly defined dry-land farming as "the attempt to store up in the soil every bit of moisture which fell and, when sufficient [moisture] had been accumulated, to call it from the ground to raise a crop."[2] Dry-land farming did not mean cultivation without water but rather cultivation without irrigation. The term itself, as well as specialized equipment, may have seemed new at the turn of the twentieth century. Contemporaries acknowledged, however, that dry-land farming had been practiced in some form since classical antiquity; they attributed its revival in the eighteenth century to the Englishman Jethro Tull through his immensely influential handbook, *Horse-hoeing Husbandry, or an Essay on the Principles of Tillage and Vegetation* (London, 1733), which emphasized deep plowing and frequent cultivation.[3]

Because the principles of dry-land farming underlie virtually every technique of horticulture on the High Plains, including what we now call xeriscaping, it is worth considering how those principles manifest themselves. To begin with, we all recognize that when it rains, some water is absorbed into the soil and some water may run off, depending on the intensity of the rain and the rate at which the soil can absorb water. Entering the soil, water will percolate far down if the soil is very sandy and not far down if the soil is very clayey. Conversely, when the hot sun beats down on the soil surface, rapid evaporation takes place. Moisture rises out of the soil, sometimes from considerable depths if arid conditions persist over a long period of time. On the High Plains, evaporation occurs more quickly when the sun's intensity combines with strong winds. Thus, without irrigation or dry-land cultivation or a combination of both, our yards, gardens, orchards, and fields suffer.

While several different forces, including gravity, act upon the movement of water in soil, of greatest interest to the dry-land cultivator are the surface forces that cause water to rise or recede. Taken together, these forces are said to operate according to the principles of capillarity. The challenge to the cultivator, then, is to use those principles, in the words of Walter Prescott Webb, "to bring the subsurface water up within easy reach of plant roots and then to hold it there and prevent its escape into the air by evaporation."[4]

In practice, Webb explained that the dry-land cultivator seeks to "increase capillarity and raise the water by initial deep-plowing, which loosens and fines the soil, and then by firming, or compacting, the soil by means of a subsurface packer[,] he can prevent or retard evaporation by creating what he calls the dust mulch. The implement most useful for forming this mulch is the harrow in some form or the other. The dust mulch acts as a blanket spread over the surface to keep the moisture from passing through the ground into the air. It is like the cork in the jug."[5]

Although the United States Department of Agriculture (USDA) Division of Botany under George Vasey, an expert on grasses and sometime explorer with John Wesley Powell, had been conducting dry-land experiments near Garden City (elevation 2,950 feet) in southwestern Kansas at least since 1889, dry-land cultivation achieved notoriety in the late 1890s through the work of an unabashed self-promoter named Hardy W. Campbell (1850–1937). A Vermonter, self-educated, and with no farming background, he had emigrated to Dakota Territory where, as a result of tinkering, he built and patented the Campbell sub-soil packer, a harrow that compacted soil at the bottom of furrows while pulverizing topsoil.

Campbell eventually settled in Lincoln, Nebraska. By 1895 he had organized his supporters into the Western Agricultural Improvement Society and started a journal called *Western Soil Culture*. In 1902 he published *Campbell's Soil Culture Manual* (revised editions in 1905, 1907, 1914) under the motto "the camel for the Sahara Desert; the Campbell Method for the American Desert." He promoted the "Campbell Correspondence School of Soil Culture" (1907–1914) and secured railroad sponsorship of extensive lecture tours throughout the High Plains.

Both the Burlington and Missouri and the Union Pacific railroads contracted with Campbell to manage "experimental" farms in western Nebraska, eastern Colorado, and northwestern Kansas.[6] These farms, as well as cooperating independent farms, concentrated on wheat, but some grew potatoes and other vegetable crops, and a few even supported dry-land fruit culture. At Parker, southeast of Denver, E. R. Parsons used the "Campbell Method" for growing cherries, plums, apples, and currants. At the Pomeroy Model Farm near Hill City (Graham County) in northwestern Kansas, Campbell oversaw the planting and maintenance of shade, ornamental, and fruit-tree seedlings over a period of five years, 1900–1904. He had been attracted to Hill City by James P. Pomeroy, a wealthy financier and western land developer associated with the railroads.

At the 490-acre Pomeroy Model Farm, Campbell took charge of preparing the soil as if for wheat: double disking, then deep plowing to eight inches, followed by subsurface packing. Seedlings were placed in holes large enough so roots could be spread out to their full lengths. Pulverized soil was packed around the roots, and a single quart of water was applied to each tree. "By vibrating the tree slightly the water soon percolated through the moist soil, dissolving the particles and settling them closely around the roots. The holes were then filled within two inches of the top, and then tramped firmly. Then about three inches of loose dirt was scattered over this packed soil, and the tree [was] left." Immediately following planting, the area was again double disked, this time to relieve the tramping of the soil by workers and wagons. After each rain, when the surface became dry, a harrow was used to pulverize the surface soil into dust mulch; and in late fall, more double disking encouraged capillary action yet still provided for seedlings to desiccate in preparation for winter. While the disposition of the Hill City farm is obscure—the farm changed ownership around 1908—we do know that Pomeroy contracted with Campbell to supervise cultivation of his farms in Colorado, Wyoming, Arizona, and Texas, although their successes are unknown as well.[7]

Despite his immense popularity, Campbell did have critics. Some argued that his "scientific soil culture" was misleading in suggesting that he had invented cultivation using limited moisture. His notion that specialized training was required turned out to be equally specious, and his technique of subsoil packing was exposed as little more than a fad. "Given a reasonable amount of common sense and close observation," wrote a farmer from eastern Colorado, "watching carefully how everything grows under different treatment and one has within him what should make a successful farmer." Other critics questioned Campbell's motives for plowing up grasslands, since he accepted financial support from the railroads and other developers. Delegates to the Dry-Farming Congress, which Campbell had organized in 1907, turned against him a year later, questioning both his cultivating techniques and the utility of his patented sub-soil packer.[8]

In western Nebraska, meanwhile, Jules Sandoz (1862–1928), a crusty pioneer turned experimental horticulturist and respected member of the Nebraska Horticultural Society, had been growing fruit trees, small fruits, and vegetables under dry-land conditions since the mid-1880s. A dropout from medical school in his native Switzerland, Sandoz took the greatest pride in successfully crossing the native plum (probably *Prunus angustifolia* Marsh.) with cultivars from Crete [Nebraska] Nurseries. His cross produced a hardy, disease-free, tender-skinned, and flavorful variety known as the Sandoz plum, no longer commercially available. He also grew sour cherries, apples, and, between rows of fruit trees, small fruit to help bind the soil and catch the snow. His daughter reminisced that every spring Sandoz gave away "wagon-loads of shrubbery, sucker plums, asparagus, horseradish, and pie-plant roots [rhubarb] to anyone who would promise to care for them." From 1903 until 1921, he reported regularly on the progress of his orchard to the Nebraska Horticultural Society. In 1910 he told the Dry-Farming Congress meeting in Salt Lake City that he had been farming for twenty-six years without irrigation and that he was cultivating 6,000 fruit-bearing trees according to the "Campbell Method." The Nebraska Agricultural Experiment Station recognized his work on behalf of horticulture on the High Plains by designating his farm southeast of Hay Springs in Sheridan County as an official experiment substation.[9]

To encourage fruit growing under similar conditions in eastern Wyoming, Aven Nelson had publicized the success of orchards in western Nebraska. His information came not from Sandoz but from E. F. Stephens, owner of Crete Nurseries and prominent member of the Nebraska Horticultural Society. Stephens described orchards along the North Platte River, near the Wyoming state line, where cultivators did use irrigation, although sparingly, in the latter part of the growing season.[10]

In 1907 the Wyoming Legislature appropriated $5,000 for dry-land farming experiments, and in 1911 it appropriated an additional $2,000 to employ a dry-land director under a newly created State Board of Farm Commissioners. The intent was to experiment not only with grains and forage crops but also with shade, ornamental, and fruit-bearing trees.[11] While the locations, duration, and success of the Wyoming experiments are unclear, they likely coincided with the establishment of dry-land experimental farms by the USDA. In 1905, the Office of Experiment Stations had established farms at Cheyenne, Newcastle in northeastern Wyoming, and Eads in southeastern Colorado.

Sandoz orchard, Sandhills, Nebraska. Courtesy, Nebraska State Historical Society.

Their purpose was to test the value of irrigating small areas, primarily for horticulture, in connection with cultivating far larger areas in row crops without irrigation—essentially, to confirm the observations of individuals such as James E. Payne in Colorado and Erwin H. Barbour in Nebraska, namely, that for security and survival on the High Plains, every farm and ranch family needed to cultivate its own garden.

The Cheyenne farm, located two miles southeast of the city, was under the direct charge of John Gordon of lower Horse Creek, Aven Nelson's correspondent. The USDA farm consisted of twenty-eight acres along the north and forty acres on the south side of Crow Creek. The north farm was divided into one-acre tracts so that, for comparative purposes, the same field crops could be grown with or without some irrigation. The source of water was a small reservoir filled by two windmill-driven wells. Downstream from the reservoir, Gordon cultivated a small irrigated plot of fruit trees, grapes, berries, and a vegetable garden; adjacent to that area was another plot with similar plants but no irrigation. In addition, he planted windbreaks, small beds of flowers, and a small lawn.[12]

Based on the 1909 growing season, Gordon concluded that "water is absolutely essential to a productive garden; in fact, such products require a great deal more water than field crops." He continued: "[T]he majority of plants are tender and do not root deeply, and to promote good growth they must be supplied with ample food, and to make this available abundant moisture is required. The irrigated garden was the admiration of all who visited it, and, considering the elevation of 6,000 feet, the results were marvelous." Except for not applying irrigation, Gordon cultivated the dry-land garden exactly as he did the irrigated garden, with results so poor as not to justify further expenditure of labor and seed.[13]

Gordon planted the Cheyenne south farm in grains, alfalfa, and potatoes, but he only irrigated during the winter with unappropriated water from Crow Creek. Potatoes proved the most successful, especially those planted in wide rows (fifty-four inches). The quality of dry-farmed potatoes was superior, so higher prices tended to offset smaller yields. The early varieties—most notably early Ohio, actually Vermont-bred in 1871 and still available through specialty houses—seemed to do best because they matured most quickly, taking advantage of spring moisture and avoiding summer heat.[14]

While the Cheyenne farm, thanks to Crow Creek, provided for comparative tests of dry-land and irrigated horticulture, conditions at Cheyenne Wells, Colorado, allowed only for dry-land activities. In 1893, as part of its campaign to convince settlers that vegetables, fruits, and trees could thrive without irrigation, the Kansas Pacific Railroad had donated a 160-acre plot, half a mile southwest of the Cheyenne Wells station, to Colorado Agricultural College for a demonstration farm.

The college named J. B. Robertson the first superintendent. When he arrived at Cheyenne Wells in November 1893, he found the quarter-section plot covered by native short-grass, except for forty acres that had been "sod busted" and tilled to a depth of four inches; the entire property was surrounded by a four-strand barbed wire fence. Thanks to a onetime state appropriation of $2,500 combined with $1,200 from Cheyenne County, Robertson built a residence and a barn, secured a team of draft horses, and purchased implements and tools. He spent the spring of 1894 disking and harrowing the forty-acre plot. His seed supply arrived late, after the soil had dried out, so germination took place only after a cloudburst provided half an inch of rain on May 30. Between then and the next storm on July 31, the heat and wind had withered most of his plants beyond recovery. Despite the fact that precipitation for 1894 amounted to less than seven inches, most of his fruit stock survived into the next spring, which was warm and dry, followed by a cool, wet summer. On

July 10, 1895, four inches of hail fell within thirty minutes, destroying most of Robertson's vegetable garden, row crops, and some tree seedlings. What he managed to save or replant was ravaged by grasshoppers later that summer.[15]

Such early horticultural failures did not appear to discourage Payne, who took over the superintendency in April 1896. In his first annual report, he expressed the view that three years was not long enough to demonstrate with any certainty the potential for agriculture and horticulture in the "Rainbelt." His perspective undoubtedly stemmed from special training in soils and problems related to root development he had received while earning a degree from Kansas State Agricultural College.[16]

Within a month of his arrival at Cheyenne Wells, Payne planted seeds of a variety of early-season vegetables. While onions and radishes were devoured by insects, his beans, lettuce, and peas survived both a dust storm (May 5–6) and a hailstorm (May 30), producing a respectable harvest. In mid-June he planted unidentified varieties of melons, pumpkins, and squashes, but only the squashes matured.

Despite Robertson's earlier failure to grow potatoes to maturity, Payne tried again, this time with four varieties including early rose, parent of early Ohio. He set out some seed potatoes on April 30 and the remainder on May 22. While Robertson's potatoes produced no tubers because of hot, desiccating winds, Payne's potatoes were attacked by beetles as soon as the vines appeared. First came the Colorado potato beetles, which Payne tried to destroy by hand-picking and crushing the mature insects, their eggs, and their larvae. In time, a "friendly bug," not identified, checked the infestation by eating larvae. Then came the blister beetles, which Payne tried to destroy with a popular insecticide known as Paris Green combined with lime or water. This remedy proved only temporary, with vines covered by beetles soon after application. "Finally, in sheer desperation," Payne reported, "we went over the field systematically every few days and killed the beetles with staves. In this way some of the vines were saved alive until the rain and hail storm of August 21." After that, the beetles disappeared, the surviving vines put on new foliage, and some tubers developed. Although Payne did not mention the size of the harvest, his first efforts at potato cultivation had proven uneconomical.[17]

With no severe storms and some well-timed rain showers during his second growing season, Payne successfully grew the Carman #1 variety of Irish potato, which proved resistant to the continuing plague of beetles. He also harvested unspecified varieties of squash, cucumber, beans, and peas. His most notable success came with the fruit trees his predecessor had planted, of which about 80 percent had survived. They included unidentified varieties of apple, cherry, plum, and "Russian" apricot. Although they had yet to produce fruit, they did show good plant growth.[18]

In May 1897, Payne set out 3,000 "Russian mulberry" saplings (probably *Morus alba* L.) from an unknown source to form a windbreak of two rows twelve feet apart, with each pair of trees two feet apart, on the north side of his forty cultivated acres. He had also received cuttings of "Russian artemesia" (probably *Artemisia glauca* Pall. ex Willd.) from Niels Hansen at South Dakota State Agricultural College. Payne had searched for a hardy, small tree to break up wind flow, reduce evaporation, and still allow for crops to grow nearby. "Some people claim," he wrote, "that I am wild on the subject when I tell them that we must either introduce systems of wind breaks here or else turn the country over to the stockmen again." In his judgment, the Russian-olive (*Elaeagnus angustifolia* L.) made the best windbreak on the arid plains, for the reasons noted earlier and because its roots, like clover, were effective in producing nitrogen for the soil.[19]

By Payne's third and last season, he reported on planting corn stalks between rows of young trees and leaving them intact to protect the trees from wind and capture winter moisture. Apparently, he had tried this approach using small fruits such as raspberries, gooseberries, and currants between rows of vegetables; however, the small fruits did not survive. Overall, efforts to demonstrate the practicability of growing horticultural products without irrigation at the substation proved unsuccessful.[20]

As an employee of Colorado Agricultural College, Payne had to know from the beginning of his stay at Cheyenne Wells that the substation operated in a financially tenuous situation. In the summer of 1895, Alfred E. True, director of the Office of Experiment Stations, and Assistant Secretary of Agriculture Charles W. Dabney had visited the Fort Collins campus to voice their opposition to using substations as demonstration farms rather than as research facilities. Dabney, former director of the agricultural experiment station and then president of the University of Tennessee, sharply criticized those who "failed to appreciate the fact that the stations are primarily scientific institutions, and that while they should always keep steadily in view the practical results to be obtained, they render the most permanent benefits to agriculture when they make thorough scientific investigations of problems underlying successful agriculture and horticulture."[21] Again, we are reminded of Charles Bessey's opinion on the proper role of the land-grant college in supporting agriculture and horticulture.

The Colorado Agricultural Experiment Station annual report for 1896 repeated Director True's view that the Hatch Act specifically prohibited the use of federal funds for substations as demonstration farms. In support of that view, the Colorado attorney general opined that applying Hatch Act monies to anything other than scientific research constituted misuse of federal funds and was thus a violation of federal law. If Colorado wanted substations, the state needed to provide the funding, something, as in Wyoming, the legislature declined to do.[22]

Strong support for Payne personally, expressed by individual members of the State Board of Agriculture, the governing board of Colorado Agricultural College, enabled him to survive the closure of the Cheyenne Wells substation and to remain on the college payroll with the title "plains field agent." As Payne later explained, his plan of work changed from studying the tests of a single person (himself) to the more valuable plan of discovering the methods of the farmers who had gained a successful foothold on the eastern Colorado plains.[23]

As noted, for three summers beginning in 1900, Payne traveled throughout eastern Colorado, primarily over what today constitutes Kit Carson, Washington, and Yuma counties. He interviewed settlers, examined their land, and took photographs. Overall, he recorded a country quite thickly resettled in the late 1880s and then dramatically depopulated by 1900. He named several settlements that had aspired to become towns, county seats, or railroad centers, for which only remnants of civilization remained: a few cellars, an occasional store building converted to residential use, but mostly foundation excavations or mounds of earth. And although thousands of preemption claims had been planted, in theory at least, as a result of the Timber Culture Act of 1873, he made a point of visiting the few groves he had heard were still in good condition and cared for. He also understood that when the country was first settled, hundreds of dry-land orchards had been planted; however, poor nursery stock, lack of experience with arid conditions, and drought had reduced the number of successful orchards to a handful.[24]

Payne realized that in eastern Colorado one could not survive, much less thrive, on dry-land farming alone. Toward that end, seven miles northeast of Flagler he discovered what he considered the foremost example of the use of topography, limited water, and human ingenuity for successful horticulture. James Howell's spread sat mostly on clayey "hard pan," cut by a small ravine, or arroyo, that ran from northwest to southeast and contained an intermittent stream. To the northwest, Howell had set out black locust (*Robinia pseudoacacia* L.), a non-native tree popular with settlers because it grows rapidly to form dense windbreaks, either in clumps or hedges. To the southeast, he had dug a shallow well to supplement limited irrigation from a small storm-water reservoir. With the small ravine dammed upstream, any water still flowing came through an artificial channel. Within the streambed, as well as on its banks, Howell had planted fruit trees—cherry, walnut, peach, and apple. Just above each tree along the bank, he had dug a basin to collect water running down during rainstorms. Payne saw some apples, and he also recorded fair yields of unspecified grapes, plums, strawberries, and gooseberries. Although apples and peaches generally required some irrigation, he concluded that cherries, plums, currants, and gooseberries could produce moderate harvests under strictly dry-land farming conditions.[25]

Undoubtedly, Howell's place had an irrigated garden somewhere near the streambed. Payne, however, chose to describe the garden at the Eckert place, about four miles southeast of the Thurman town site (halfway between Flagler and Anton in present-day Washington County). Peter Eckert had dug two wells, each 112 feet deep, and used windmills to pump water supplies into storage reservoirs. By operating continually, the pumps provided enough water to irrigate two acres and supply his cattle. The homestead sat on a rise, so Eckert had terraced his irrigated plot for the most efficient use of water. He divided the plot into sections for ornamental plants, a vegetable garden, an orchard excluding apples, an apple orchard, and a grape arbor. Payne learned that while hail and wind had prevented the normal harvest of fruit, Eckert still kept the orchards in excellent condition. No longer able to do the heavy work of dry-land farming, Eckert devoted his strength to gardens and trees.[26]

After his first summer traveling through eastern Colorado, Payne reached the same conclusion as Nebraska's Erwin H. Barbour who, as noted, had argued that to survive and sustain themselves, isolated homesteaders on the High Plains required not only protection from the wind but at least minimal irrigation, enough to support a vegetable garden, a few fruit trees, and some livestock. For those seeking to cultivate gardens and orchards for family use, Payne advised that they could succeed by selecting favorable locations "and then working hard with both hands and intellects." Overall, Payne concluded that the plains of eastern Colorado lent themselves best to raising livestock and not, as we see today, to cultivation of grain and forage crops.[27]

By 1903, seven years after his arrival in eastern Colorado, Payne could report visible progress in civilizing. He had seen sod houses replaced by frame houses and noted other permanent improvements. He interpreted seeing fewer places with an "I want to sell" appearance to mean that people had decided to put down roots and create homes.[28]

The circumstance of Payne's departure from Colorado Agricultural College and his employment by the newly created USDA Office of Dry-Land

Windbreak demonstration, Akron, Colorado, 1927. Courtesy, Colorado State University Archives and Special Collections, Cooperative Extension Collection.

Agriculture is obscure. In 1907 he headed the dry-land experiment station at Garden City, formerly operated by the Division of Botany, but he soon returned to eastern Colorado as first superintendent (1908–1910) of the new dry-land experiment station at Akron. Both stations were among the twenty-four reorganized or established by E. C. Chilcott, agriculturalist in charge of the Office of Dry-Land Agriculture.

Akron's initial horticultural distinction came from research on shelterbelts, the narrow strips of trees planted as windbreaks to protect homesteads, crops, and livestock, with results that favored the ponderosa among conifers and the hackberry (*Celtis occidentalis* L.) among deciduous trees. In addition, Akron served as the site for early experimental work on minimal water requirements to produce grain and forage crops, with results eventually applied to horticulture. Researchers demonstrated the fallacy of any single "system" of dry-land farming, most notably the "Campbell Method." Their observations and experiments illustrated that deep plowing did not invariably increase the water-holding capacity of the soil and that dust mulch was not necessarily the best way to preserve water in the soil after initial cultivation.[29]

The limits of dry-land horticulture, which Payne observed on the plains of eastern Colorado, had been recognized since the mid-1870s in western Kansas. To be sure, that part of the High Plains had been settled earlier than

Learning to plant ponderosa pine, Washington County, Colorado, 1932. Courtesy, Colorado State University Archives and Special Collections, Cooperative Extension Collection.

eastern Colorado; the climate is somewhat less arid, and the elevation is lower by 1,000 to 2,500 feet. By trial and error, settlers in western Kansas had concluded that market gardening would not work but that home gardening could succeed with hard work, intelligence, and, where possible, some irrigation.[30]

W. D. Street of Oberlin attributed the relative neglect of home gardening and orchards in western Kansas to the overwhelming attention given to production of wheat and corn. Frequent dry spells, accompanied by high winds and the rapid growth of weeds, added to the discouragement. After several years of trial and error, Street adopted three rules for successful home gardening. First, plant early, and then only with standard "cold-season" varieties of vegetables; when possible, successively replant with those same varieties: "Our experience in a climate of some uncertainty, and without irrigation, has taught us that it is best to get the crop grown to maturity as rapidly as possible, which reduces the chances of failure very materially, and, in many instances, saves the crop." Second, use plenty of well-rotted manure, plowed in deeply in the fall and plowed as top dressing in the spring. Even with rich soil, such application helps retain soil moisture. Third, devise a simple, economical way to preserve water for year-round usage. Street had built a small dam with headgates and flumes that held a reservoir of water sufficient for garden, orchard, and farm. As a result, the yield from just a few irrigated acres greatly exceeded that of many additional nonirrigated acres.[31]

Typical of the "practical guides" written for specific localities throughout the High Plains, Street listed the vegetable varieties he found did best season after season. Among them, some are still available: golden wax beans, early Jersey wakefield cabbage, black-seeded simpson lettuce, and early Ohio potato. In addition, Street listed without indicating varieties: asparagus, beets, carrots, cucumbers, horseradish, muskmelons, onions, peas, pumpkins, radishes, spinach, squashes, tomatoes, and turnips. Altogether, he believed northwestern Kansas could produce some of the best gardens and orchards in the state.[32]

Concerning shade trees and windbreaks, W. A. Mikesell of Atwood, also in northwestern Kansas, found the ash (*Fraxinus americana* L. is native) and black walnut (*Juglans nigra* L., native to central Kansas) the most hardy, cottonwood and boxelder the fastest growing, and black locust (*Robinia pseudoacacia* L.), like other imports from eastern Kansas, satisfactory if properly cultivated. He advocated deep plowing and pulverizing of the soil in the fall prior to planting and cultivation to control weeds, at least until groves or windbreaks were well established. If cuttings were to be used, they must be made in fall before hard frost, kept moist but not frozen over the winter, and planted in spring when the weather turned favorable for planting corn. Otherwise, seeds could be collected, kept dry, then planted with a light cover of soil in late fall or early spring.[33]

At least since 1878, residents of western Kansans had recognized that for horticulture, and agriculture generally, to thrive, the state needed to support "men of science" financially in their work on plants under field conditions. In other words, research useful to western Kansas could not be conducted at the agricultural experiment station at Kansas Agricultural College in Manhattan.[34]

Fortuitously, closure of the Fort Hays Military Reservation, consisting of 1,400 acres of bottomland along Big Creek and 2,200 acres of gently rolling upland, provided the opportunity for the establishment of the first state institution in western Kansas. In February 1895, State Representative J. Schlyer of Ellis County (county seat: Hays) introduced a resolution to the legislature petitioning Congress to transfer the reservation from the United States to the state of Kansas for the purposes of establishing a branch agricultural experiment station, preserving a grove of native trees as a public park, and converting existing military buildings into the western branch campus of the Kansas State Normal School at Emporia. Five years later Congress approved, and President William McKinley signed the requested legislation. In January 1901, the legislature and Governor W. E. Stanley entrusted the former reservation to the regents of the Agricultural College and the Normal School. Since federal monies could no longer be used for experiment substations, the legislature appropriated $3,000 annually for the years 1902 and 1903, enough to make the Hays experiment station operational.[35]

While dry-land research on grain and forage crops dominated work at the station (now known as the Western Kansas Agricultural Research Center), some horticultural demonstration activities did occur during its early years. Beginning in 1902, J. G. Haney, the first station superintendent and a forage expert, instructed that an area formerly used by the military as a garden be planted with vegetables including unspecified varieties of potato, cabbage, beans, corn, tomato, radish, lettuce, carrot, parsnip, red beet—and Jerusalem artichoke (*Helianthus tuberosus* L.), a virtually indestructible perennial sunflower that produces edible subterranean potato-like tubers that taste somewhat like artichokes. Because of an unusually wet growing season, the first garden did well without irrigation. Subsequent gardens received some irrigation from wells built to raise subterranean water levels of alfalfa fields. By 1909, when these experiments in sub-irrigation had proven ineffectual and were discontinued, irrigation water for horticulture came directly from Big Creek.[36]

Experiment station, Hays, Kansas, ca. 1920. Courtesy, Western Kansas Agricultural Research Center.

Early attempts at fruit culture were only slightly more successful. The first fruit trees, shipped from eastern Kansas, were planted during the spring of 1903. Within three years the orchard had expanded to nine acres, containing about 1,300 specimens of unspecified varieties of apple, crab apple, peach, pear, plum, cherry, and apricot. Haney and staff documented yields, survival rates, and pruning practices but conducted no real research. The usual high winds and late-spring frosts appeared the leading obstacles to the development of a viable orchard. As fruit trees died, they were replaced primarily with the early Richmond and Montmorency cultivars of sour cherry (*Prunus cerasus* L.), the same cultivars that became popular along Colorado's Front Range. In addition, during the early years, several varieties of grapes, strawberries, blackberries, and raspberries were planted; the grapes in particular succumbed to winter kill.[37]

Hays's major contribution to western Kansas horticulture was as nursery for the propagation and distribution of shade trees and windbreaks. Beginning in 1903, the station purchased both conifer and deciduous stock from federal and commercial nurseries in the Midwest and California. Within two years, inventory consisted of about 15,000 trees—Austrian, Scotch, and ponderosa pine, cedar (likely *Juniperus virginiana* L.), redbud (*Cercis canadensis* L.), bur

Harvesting cherries, experiment station, Hays, Kansas, 1937. Courtesy, Western Kansas Agricultural Research Center.

oak, black walnut, honey locust, Kentucky coffeetree (*Gymnocladus dioica* (L.) C. Koch), white elm (*Ulmus americana* L.), red elm (*Ulmus rubra* Muhlemb.), hackberry, persimmon (*Diospyros virginiana* L.), green ash, Russian mulberry, Russian-olive, various poplars, and maples. It is unclear whether cuttings were ever taken from native stock either at the public park along Big Creek or elsewhere in western Kansas. The station does possess a typed checklist of native trees and shrubs prepared in 1903 by Royal Kellogg, the forester who had earlier participated in the survey for the Nebraska National Forest, as well as his article on how to plant forest trees in western Kansas.[38]

Again, horticulture was ancillary to the station's primary mission: to serve stock farmers. In 1907, under the direction of C. K. McClelland from the USDA Office of Farm Management, the first shelterbelt was planted on a sixty-acre plot to demonstrate the value of windbreaks in protecting orchards. Inside the shelterbelt, thirty acres were planted to 1,000 fruit saplings (varieties unknown); the shelterbelt itself, around 160 feet wide, consisted of 17,000 catalpa (probably *Catalpa bignonioides* Walt.), honey locust, and Osage-orange seedlings. An exceptionally severe drought during the first growing season killed most of the stock, and the plot was entirely replanted. Additional seedlings were set out in the spring of 1909; a hailstorm that summer destroyed 97 percent of the fruit trees and severely damaged the windbreak trees. This time, the orchard was not replanted. Lacking proper maintenance, the shelterbelt deteriorated and eventually was uprooted.[39]

In recognition of the need to plant more trees on Kansas farms, the 1910 legislature created a Division of Forestry within the Agricultural College and entrusted the regents to appoint a state forester over "all experimental and demonstration work in forestry conducted by the Experiment Station." The regents named Charles A. Scott, a Kansas Agricultural College graduate who had also participated in the Nebraska survey, as the first Kansas state forester. In appointing Scott, the regents directed that nursery stock be grown at Hays as well as Manhattan and that such stock be made available at cost to any Kansas resident. Scott appointed J. F. Brandon nurseryman in charge of the Hays nursery. During 1912, Brandon distributed 67,082 trees throughout western Kansas: red cedar, catalpa, green ash, white elm, and honey locust. Based on responses from recipients, who were required to report back to the station as part of their agreement in accepting nursery stock free of charge, three-quarters of the trees were still alive after three years, despite unusually poor growing seasons.[40]

Arid climate may have been the principal hindrance to the progress of forestry, and horticulture generally, in western Kansas. In addition, another obstacle received at least as much attention: jackrabbits. Battling jackrabbits continues to be among the most frustrating and least successful endeavors on farms and in rural communities throughout the High Plains. Actually, jackrabbits are not rabbits but large hares of the genus *Lepus*. Their very long, donkey-like ears earned them the name "jackass rabbits," later shortened for respectability; their long hind legs allow them to run with amazing speed. Because of their size (up to ten pounds) and their rapid expenditure of energy, jackrabbits are voracious eaters, giving credence to the adage that one jackrabbit eats as much as a horse.

Hays's first superintendent documented the damage caused by jackrabbits, eating farm and garden vegetables and, equally destructive, gnawing the bark off young fruit trees. J. G. Haney described a protective device invented by his farm crew: a thinly sawed piece of wood that, when wet, could be bent around tree trunks, then fastened by wire—an expensive, time-consuming activity impractical to duplicate on farm and homesteads. Horticulturists at the Kansas Agricultural Experiment Station, as well as those at other High Plains stations, did their best to devise additional preventative measures and protective treatments. For the former, the most obvious recommendation was

Rabbit damage to apple tree, Cheyenne Horticultural Field Station. Courtesy, USDA–Agricultural Research Service.

to remove brush piles and thickets, the jackrabbits' breeding and hiding places. For the latter, the Kansans provided diagrams for a "novel and ingenious" trap that consisted of a wooden box with a door hung by two hinges and connected to a simple trigger mechanism. When released, the mechanism prevented the jackrabbit from getting out—similar in design to the present-day "Havahart" live-animal traps. In addition, various repellents and poisons were described, but only concoctions containing the highly toxic strychnine were guaranteed effective. Since jackrabbits seem to like fruit and sugar, one lethal concoction required the user to "chop apples or melons into small cubes. Add sugar equal to one-half the weight of the fruit. Boil until the mass forms a thick jam. Add strychnine, either powdered or dissolved, at the rate of one ounce to 26 pounds of jam, and mix thoroughly." Besides the indecent waste of jam, we have since learned that the main disadvantage of strychnine is that it kills indiscriminately.[41]

Especially in farm journals such as *Field and Farm*, much ink was devoted to the war against jackrabbits. No method, however, has received the publicity of the periodic roundups in which men and boys, and sometimes girls with

Organized jackrabbit kills, Morgan County, Colorado, 1937. Courtesy, Colorado State University Archives and Special Collections, Cooperative Extension Collection.

light weapons, would set their sights on a relatively small area such as a quarter-section, then quickly shoot dozens and sometimes even hundreds of jackrabbits, resulting in exterminating most of the species—at least for a time.[42]

Jackrabbits, of course, were not nearly as prevalent in the more urban areas of the High Plains. By the late 1880s, *Field and Farm*, an outspoken advocate of war against jackrabbits, had turned to ornamental horticulture as its own, more genteel cause, not just for aesthetic reasons but also for making homestead and community more inviting and more livable.

Notes

1. J. E. Payne, "Investigation of the Great Plains. Field Notes from Trips to Eastern Colorado," *Colorado Agricultural Experiment Station Bulletin* 59 (1900): 5–6.

2. Quoted in Everett N. Dick, *Conquering the Great American Desert: Nebraska* (Lincoln: Nebraska State Historical Society, 1975), 355–356.

3. *Field and Farm* 27, no. 1364 (March 23, 1912): 6, and 28, no. 1412 (February 22, 1913): 6.

4. Walter Prescott Webb, *The Great Plains* (1931; repr. Lincoln: University of Nebraska Press, 1981), 370–371.

5. Ibid., 371; see also K. G. Brengle, *Principles and Practices of Dryland Farming* (Boulder: Colorado Associated University Press, 1982).

6. Mary W.M. Hargreaves, "Hardy W. Campbell (1850–1937)," *Agricultural History* 32 (January 1958): 62–64; Hardy W. Campbell, *Campbell's 1907 Soil Culture Manual* (Lincoln: Campbell Soil Culture Co., 1909), 58.

7. Campbell, *Soil Culture Manual*, 206–217, 258 (quote on p. 208).

8. R. C. Nisbet, "How about This Campbell Theory?" *Field and Farm* 22, no. 1114 (May 4, 1907): 4; Hargreaves, "Hardy W. Campbell," 64–65.

9. Mari Sandoz, *Old Jules: Portrait of a Pioneer* (1935; repr. New York: MJF Books, 1996), 248; Jules Sandoz, "Plums for the Northwest," Nebraska State Horticultural Society Annual Report 38 (1907): 189; *Dry-Farming Congress Bulletin*, cited in Dick, *Conquering the Great American Desert*, 371.

10. Aven Nelson, "Seasonal Suggestions," *Ranchman's Reminder* 5, no. 5 (May 1908): 42; E. F. Stephens [communication to Aven Nelson], *Wyoming State Board of Horticulture Special Bulletin* 2 (1910): 36.

11. Frances Birkhead Beard, *Wyoming from Territorial Days to the Present* (Chicago: American Historical Society, 1933) 2, 559; *Session Laws of the State of Wyoming* (1911), Ch. 88, Sec. 1–4: 134.

12. O. W. Bryant, "Progress Report on Experiments in Supplemental Irrigation with Small Water Supplies at Cheyenne and Newcastle, Wyoming, 1905–1909," USDA Office of Experiment Stations Circular 92 (Washington, D.C.) (January 1910): 6–7, 12.

13. John H. Gordon, "Experiments in Supplemental Irrigation with Small Water Supplies at Cheyenne, Wyoming in 1909," USDA Office of Experiment Stations Circular 95 (Washington, D.C.) (April 1910): 9.

14. Bryant, "Progress Report," 29.

15. Colorado State Agricultural Experiment Station Annual Report (1894): 9, 19, 31, 70, 73, and (1895): 137, 141–142.

16. Ibid. (1896): 94; Alvin T. Steinel, *History of Agriculture in Colorado* (Fort Collins: Colorado State Agricultural College, 1926), 276.

17. Colorado Agricultural Experiment Station Annual Report (1896): 180.

18. Ibid. (1897): 100–106, and (1896): 109.

19. Ibid. (1897): 106; quote from J. E. Payne, "Wild Olive," *Field and Farm* 13, no. 654 (July 16, 1898): 3.

20. J. E. Payne, "Windbreaks and Hedges," Colorado State Board of Horticulture Annual Report (1898): 72–74; Payne, "Wild Olive," *Field and Farm* 13, no. 676 (December 17, 1898): 3.

21. Colorado Agriculture Experiment Station Annual Report (1895): 90.

22. Ibid. (1896): 88–89.

23. Ibid., 217; Payne, "Field Notes," 3.

24. Payne, "Field Notes," 4–8.

25. Ibid., 7–9.

26. Ibid., 9.

27. Ibid., 16; J. E. Payne, "Irrigation of Orchards and Gardens under Difficulties," Colorado State Board of Horticulture Annual Report (1899): 40.

28. J. E. Payne, "Unirrigated Lands of Eastern Colorado: Based on a Study and Residence of Seven Years," *Colorado Agricultural Experiment Station Bulletin* 77 (1903): 14–15.

29. Bentley W. Greb and D. W. Robertson, "Fifty Years of Agricultural Progress, 1907–1957, USDA, Colorado State University, and Local Cooperating Groups, June 28, 1957" (Akron, Colo.: USDA Central Great Plains Field Station, 1957): 1–2; Lyman J. Briggs and H. L. Shantz, "The Water Requirements of Plants. I: Investigations in the Great Plains in 1910 and 1911," USDA *Bureau of Plant Industry Bulletin* 284 (Washington, D.C.) (1913): 3; Steinel, *Agriculture in Colorado*, 559–562.

30. S. C. Mason and F. C. Sears, "Small Fruit Culture by Irrigation," *Kansas Agricultural Experiment Station Bulletin* 55 (December 1895): 1.

31. W. D. Street, "Vegetable Gardening in Northwest Kansas," Kansas State Horticultural Society Annual Report 26 (1896): 61.

32. Ibid., 60–63.

33. W. A. Mikesell, "Forestry in Rawlins County," Kansas State Horticultural Society Annual Report 16 (1886): 128.

34. Martin Allen, "Horticulture on the Plains," Kansas State Horticultural Society Annual Report 8 (1878): 187–189.

35. Leland E. Call and Louis C. Aicher, "A History of the Fort Hays (Kansas) Branch Experiment Station 1901–1962," *Kansas Agricultural Experiment Station Bulletin* 453 (May 1963): 3–8, 12.

36. Ibid., 32–35.

37. Ibid., 40–42; Fort Hays Agricultural Experiment Station Annual Report (1952): 35, typescript; J. G. Haney, "Experiments at the Fort Hays Branch Station, 1902–04," *Kansas Agricultural Experiment Station Bulletin* 128 (December 1904): 278–279, 303, 316–318.

38. Call and Aicher, "Fort Hays," 36; see also Royal S. Kellogg, "Notes on the Native Woody Species of Western Kansas," 42 p., typescript (Hays: Western Kansas Agricultural Research Center, 1903).

39. Call and Aicher, "Fort Hays," 37.

40. Raymond J. Pool, "Fifty Years on the Nebraska National Forest," *Nebraska History* 34 (September 1953): 146–147; Charles A. Scott, "Provisions of the State Forest Law (1909)," Kansas Agricultural Experiment Station Circular 10 (1910): 1–3 (quote on p. 1); Scott, "Trees for Kansas," Kansas Agricultural Experiment Station Circular 55 (January 1916): 17.

41. Haney, "Experiments at Fort Hays," 303; J. C. Cunningham, "Protecting Trees from Rabbits," Kansas Agricultural Experiment Station Circular 17 (February 1911): 1–4 (quote on p. 4).

42. See, for example, A. C. Covert's letter to the editor, *Kansas Farmer* 9, no. 9 (May 1, 1872): 139; "War on Rabbits," *Field and Farm* 16, no. 798 (April 20, 1901): 4; C. S. Harrison in *Field and Farm* 23, no. 1149 (January 4, 1908): 3; A. Homan, "The Man with the Hoe," *Field and Farm* 28, no. 1411 (February 15, 1913): 2.

8

Forging New Paths in Ornamental Horticulture

In a year-end essay in 1906, their twentieth year in business, the publishers of *Field and Farm* editorialized on "a gradual change coming over the customs of our agricultural people so that their surroundings take on the aspect of eastern environments." By this they meant not only better-built homes and well-maintained yards and gardens but also the amenities of a "higher civilization" such as rural mail service, neighborhood telephone systems, Grange and lyceum associations, women's clubs, and farmers' institutes—all helping to reduce the drudgery of rural life as it had long been known. To be sure, those amenities by no means reached all rural homes, but they were sufficient for the publishers to conclude that the region's residents were "becoming more easternized year by year and are gradually growing out of the wild and wooly traditions of the good old pioneering days."[1]

Despite its name, *Field and Farm* not only catered to the "intelligent agriculturalists" on the isolated homesteads and in the small communities of the arid West but also appealed to city dwellers involved in the cultivation of gardens, whether as professionals or amateurs. Indeed, *Field and Farm* was headquartered in Denver, which, since the late 1880s, had distinguished itself as the principal city of the mountains and plains.

Since its founding in 1886, *Field and Farm* had editorialized in favor of gardening, tree planting, landscaping, and horticulture generally. In fact, in their first year, the editors successfully urged Governor Benjamin Eaton to proclaim Colorado's first Arbor Day, noting that Colorado's neighbors, Nebraska and Kansas, already celebrated that holiday. In his proclamation, Governor Eaton tied together planting trees for the present and putting down roots for the future, between "beautifying our homes, cemeteries, highways, public parks and landscape" and "making thoughtful provision for the happiness of those who are to come after us."[2]

Besides encouraging residents to volunteer a day to plant trees, Eaton urged public school teachers throughout Colorado to engage their students in "proper and practical observance" of Arbor Day. For the first annual celebration, *Field and Farm* reported that Denver students assembled in City Park from 10:30 until noon; each school was assigned a grove or walkway named for the occasion to honor a national or local notable, for example, Ralph Waldo Emerson and Colorado's first territorial governor, William Gilpin. Students at the various locations listened to short speeches—including readings from writings of the respective notables—and celebrated with songs, making the park "ring with the noise of their glad young voices."[3]

Celebration of Arbor Day was regularized in 1890, with the third Friday in April officially declared a holiday in all Colorado public schools. Teachers and principals were required to engage their students in the planting of trees and other civic exercises; school superintendents had to report annually to the state forester on Arbor Day activities within their respective counties.[4]

Despite the published rules, the editors of *Field and Farm* grew increasingly critical of the failure to achieve practical results. After Arbor Day 1896 passed with little school activity statewide, the editors lamented that the holiday had become "a good deal of a farce," but it could have been an "occasion of grand achievement." Turn every schoolhouse grounds into an arboretum, the editors suggested, so that students could easily study all species of trees growing in Colorado. In addition to serving as one of the most attractive sites in any community, school grounds could also provide space for gardens where children could both learn and work. By way of precedent, in France gardens existed at 30,000 public schools where primary schoolteachers were required to give practical instruction in horticulture.[5]

While the editors of *Field and Farm* promoted the narrowly practical and vocational benefits of teaching children about horticulture, Colorado's superintendent of public instruction from 1898 to 1904 held a much broader view of the function of the school garden in the curriculum. To be sure, children in rural schools should receive practical training in horticulture and agriculture. Beyond that, however, Helen Thatcher Loring Grenfell (1863–1935) envisioned the school garden as a valuable aid in teaching the basic principles of the natural sciences to all children, both rural and urban. Responsible citizenship in the modern world, she noted, required a comprehensive knowledge of the natural sciences; much can be gained in school by personal observation and personal effort that will be eminently useful later in adult life.

School gardens can have an even broader utility in providing the surroundings for children to strengthen their powers of perception. For example, frequent contact with trees and flowers arouses a love of nature in children, quickens their abilities to observe, cultivates their love for the artistic and beautiful, and renders their senses more acute and their judgments clearer. For another example, garden work instills in children a feeling of respect and sympathy for all manual labor. According to Grenfell, "Children of almost any age can take part in some way in horticulture; it is a kind of work that they enjoy, and through it a taste will be developed for labor . . . which will make them feel akin, not alone to 'the man with the hoe,' but with the world's toilers in every line, no matter how humble their work." In other words, by nurturing in children virtues such as diligence, attentiveness, judgment, and self-reliance, gardening develops moral character.[6]

Helen Grenfell was a missionary for "progressive education" before it became fashionable. She was born in Valparaiso, Chile, where her father had served in the diplomatic corps; she spent her infancy in Concord, New Hampshire (her maternal grandfather published the *Statesman* newspaper); and at age six she was brought to Central City, Colorado, then to a farm west of Longmont. Characteristic of the era, she began teaching at age sixteen, then attended teachers' college in Albany, New York, returned to teach in Longmont, and eventually served as school superintendent of Gilpin County before being elected without opposition on a multiparty, or fusion, ticket as state superintendent of public instruction in 1898.[7]

Helen Greenfell, of course, did not invent the idea of the school garden. Indeed, she credited Cyrus the Great, founder of the Persian Empire in the mid-sixth century B.C., with establishing the first school garden, where sons of courtiers received instruction in horticulture. She also noted that Jan Comenius, a seventeenth-century Moravian bishop and teacher, and, later,

Johann Heinrich Pestalozzi, a Swiss education reformer, advocated a garden for every primary school. In fact, during the second half of the nineteenth century, Sweden established primary school gardens, Belgium made horticulture a required course and school gardens universal, and in France, as noted, school gardens were attached to most elementary schools, meaning the normal schools taught agriculture and horticulture in order to prepare teachers. By the turn of the twentieth century, a few individual schools in the United States included gardens on their grounds, most notably the George Putnam School in Boston.

In Colorado, Superintendent Grenfell acknowledged that "as a rule our country school houses are located in the most arid spot in an arid region" and that, especially on the eastern plains, little could be grown without some irrigation. She expressed confidence that enough public-spirited individuals could be found to ensure water for school gardens. If they could not deliver water to school properties, perhaps farmers could donate some of their better land near the schools. Until changes occurred, most of the state's school grounds would remain "an illustration of the abomination of desolation, and their influence upon our young people is, to say the least, not of an elevating nature."[8]

Years earlier, Mrs. Arthur E. Gipson, wife of the pioneer Greeley nurseryman, had addressed the Colorado State Horticultural Society on the subject of horticulture as a means of moral education for children in urban schools. Mrs. Gipson expressed her belief that children learn patience by planting seeds and waiting for plants to emerge, learn gentleness by caring for young plants, and learn to love the beautiful by observing buds and blooms. Committed to the ornamental side of horticulture, Mrs. Gipson anticipated the time when the school flower garden would be as common as the spelling book.[9]

When the Colorado State Horticultural Society invited Helen Grenfell, then in her last year as superintendent, to address its annual convention in 1904, the membership probably did not anticipate the vigor of her discourse against cutting Christmas trees. She began by reminding her listeners that the tradition came from paganism and was not part of the original celebration of Christ's birth. Nor was the wanton destruction of trees in any way a Christian act. On the one hand, she noted, we impress our children with the usefulness and significance of trees, for example, by planting thousands of trees on Arbor Day. On the other hand, for Christmas we cut down twenty, even fifty times as many trees as we plant on Arbor Day. Surely, we can find another way to make Christmas happy for our children "without offering up as a sacrifice so many little evergreen trees." Indeed, the society was moved to pass a resolution against the cutting of Christmas trees, although, clearly, its message did not appeal to the general public.[10]

While Helen Grenfell's opposition to cutting Christmas trees may seem a bit eccentric, if not wholly impractical, her approach to the study of nature and the implications for activism were within the main line of the eighteenth-century Enlightenment—all about the "laws of nature and nature's God." Her Congregational upbringing instilled a belief in one God who created the universe but who has not intervened in its operation since the creation. Such deism naturally led to the view, shared in various ways by many of her contemporaries, that here on earth God's work must truly be our own. One of those contemporaries, Alice Eastwood (1859–1953), grew up in Colorado, taught at East Denver High School, and botanized along the Front Range before she embarked on a lengthy botanical career based at the California Academy of Sciences in San Francisco.[11] It is difficult to imagine that Helen Grenfell and Alice Eastwood, at one time both schoolteachers in the Denver area, did not know each other.

It is also difficult to imagine that they did not share Charles Bessey's view that horticulture cannot be successfully pursued without knowledge of the essentials of botany. Like Bessey who had written a general textbook for high school students and the shorter *Elementary Botanical Exercises for Public Schools and Private Study* (1894) and Aven Nelson who wrote *An Analytical Key to Some of the Common Flowering Plants of the Rocky Mountain Region* (1902) for use by public schoolteachers and their students, Eastwood published *A Popular Flora of Denver, Colorado* (1893). Written primarily for high school students, her *Flora* served as a field introduction to botany as part of general education.

Somehow, botanists had not convinced enough teachers to use their school yards and neighborhoods as instructional laboratories, nor had enough teachers made the connection between experiential learning and practical results—at least not in the view of *Field and Farm*. Just prior to Arbor Day 1902, the editors again grumbled that there would be the usual half-day holiday in the schools, some instrumental music and singing, but few if any trees planted. Those planted, moreover, would be neglected as soon as schools closed for the summer. To date, they continued, Arbor Day had served simply as an excuse for children to have fun. Instead, it should be an opportunity for teachers to cultivate in their students a genuine love of plant life, not through recitation of some literary masterpieces that are beyond the children's mental grasp but by a study of actual trees. Children should learn to observe the character of the bark, how branches are developed, and so on, and then, still under their teachers' supervision, learn to plant trees carefully and correctly.[12]

While attractive school surroundings may contribute to appreciating beauty and studying nature, preparation for learning, as we are often reminded, starts around the home. Toward that end, *Field and Farm* made its pages available to horticultural missionaries, most notably Charles S. Harrison of Nebraska who preached turning the rural homestead into "the home beautiful." Addressing the issue of young people emigrating, Harrison argued that the "home beautiful" could reverse that trend. Using anecdotal evidence, he noted that children brought up on attractive homesteads and who then left for school or jobs were the most eager and most likely to return to their home country. An unattractive homestead, on the other hand, which looks like the "front yard of a hospital for disabled machinery, with scents of the hog pen and barnyard floating around the premises, with clouds of flies, with only cells for bedrooms," conveys a feeling of "listlessness and indifference which is fatal to the future." Beautifying the homestead makes the place worth more financially, but it also increases the chance that young people will eventually return.[13]

For a relatively brief period before the turn of the twentieth century, the county fair had served as the principal vehicle by which rural residents found new ideas and practices for community, homestead, and self-improvement. By 1900, however, the county fair no longer held to its original educational purpose. According to *Field and Farm*, exhibits of local produce and goods came to illustrate the economic advantages of a given area to prospective outside investors and new settlers, while the side shows and "fake games" were meant to attract the homesteaders' dollars. Furthermore, larger regional events such as Denver's annual Festival of Mountain and Plain, publicized nationwide, were used to illustrate not only the horticultural and pastoral but also the industrial, mining, and mercantile advantages of Colorado.[14]

Just as *Field and Farm* sought to promote beautification of isolated homesteads, it also crusaded for beautifying towns and cities. "No one has a reasonable excuse," the editors wrote, "for living on a place in this new country, six, eight or ten years, perhaps, as we so often see them, without showing by their works that they have done something at least to beautify their home and make it pleasant. They owe this to their families as well as to the community in which they live."[15]

No community on the High Plains was more committed to beautification than Greeley. Within a mere twenty years of its founding on the arid plains of northeastern Colorado, Greeley could describe itself in truth as the "Garden City," thanks to irrigation. By 1890, most of Greeley's original, makeshift

8th Street, Greeley, 1890s. Courtesy, City of Greeley Museums, Permanent Collection.

homes had been replaced with permanent brick and wood-frame dwellings, surrounded by lawns as well as ornamental and vegetable gardens. By then, too, trees planted along the 100-foot-wide avenues and streets had grown enough to provide protection from sun and wind as well as to adorn the community with a certain dignity and majesty.[16]

Prominent among the founders of Greeley as a "Garden City" was an eccentric florist named John Leavy (1831–1911). Born in Dublin, Ireland, he had been preparing for the priesthood when, apparently as a result of visits to the Royal Botanical Gardens at Kew, he switched to ornamental horticulture. Immigrating to New Jersey in the early 1860s, he had been among the very first to answer Nathan Meeker's call to join the Union Colony. Once in Greeley, he built his own cabin and surrounded it with trees, shrubs, and flowering plants. He also built and maintained a greenhouse and, nearby, several plant beds heated by fermented manure from which he provided the earliest known plant specimens for Greeley's public places. His own vegetable garden became a sort of bellwether for community horticulture; each spring, local newspapers reported on what he planted, his first harvests, and his yields. Bushy-bearded, bedraggled looking, yet beloved by the community, he gained the reputation of nature lover and nature worshipper. Whether eccentricity is inseparable from utopianism can be debated, but the fact that gardening has always attracted eccentrics is certain.[17]

As to the smaller towns around Greeley and beyond, as irrigation ditches and laterals were developed along streets and alleys, they too changed character from clusters of sod houses or tent encampments to oases of permanent dwellings. Shade trees, most frequently the native cottonwood, were planted along streets and in front yards, while fruit trees were planted in backyards. Early settlers who had left these towns and returned many years later, according to *Field and Farm*, barely recognized their appearance.[18]

Beginning in the 1890s, support of community beautification came most notably from women's clubs, with the establishment of public parks among their earliest and most visible projects. In Gering, Nebraska (population 200), for example, the Women's Club secured the donation of a town block for a park; landscaped the property with trees, shrubs, and flowering plants; and ensured the park's maintenance.[19]

In larger towns, members of the Citizens Tree Committee appointed by the Cheyenne City Council, together with members of the City Improvement Association (precursor to the Chamber of Commerce), raised the money and volunteered the labor to plant several thousand saplings annually, beginning in the late 1890s. Businessman James F. Jenkins, an early organizer of tree planting and chair of the Citizens Tree Committee, reported that for the 1902 growing season, 250 adults and 350 schoolchildren raised $875 to obtain 2,000 cottonwoods and an unknown number of ash and maple seedlings and that the city had assumed responsibility for irrigation and pruning. In addition to the planting along public thoroughfares, many homeowners had purchased trees through the municipality and planted them along vacant lots in residential areas.[20]

Because tree planting in Cheyenne, as elsewhere, had occurred on an ad hoc, voluntary basis without expert assistance, saplings were often planted too close together, causing unnecessary expense and, in time, unhealthy trees, while parts of the community remained without any trees. Furthermore, citizens planted either those varieties they could find in the surroundings or stock sold to them by "tree peddlers," regardless of suitability to local conditions. In other words, Cheyenne, like most communities on the High Plains, lacked a comprehensive and practical plan for city-wide planting until the Cheyenne City Council contracted with Saco Rienk DeBoer (1883–1974) to develop such a plan in 1911. A Dutch immigrant, DeBoer had distinguished himself as the first landscape architect for the City of Denver.

After a preliminary survey of Cheyenne, DeBoer noted the obvious: as in most communities on the High Plains, cottonwoods had been planted to the exclusion of other, in many ways more satisfactory, varieties of trees. He warned of the danger inherent in planting only one variety, pointing to the black locust borer that had destroyed virtually all black locusts in Denver during the 1880s and to the brown-tailed moths and leopard moths threatening the American elm in northeastern states. He argued that planting a variety of trees could lessen, if not entirely prevent, such troubles.

DeBoer identified eighty varieties of trees and shrubs that he thought, with proper planting and maintenance, could thrive in Cheyenne. Of that number, he recommended about forty varieties for street planting and the remainder for yards. He developed a detailed, block-by-block plan for tree planting and recommended that funding be secured through special tree districts, in the same way sidewalks, curbs, and gutters were funded. Maintenance of trees along streets needed to be done by "an expert city employee" whose salary came from the tree districts. For the outlying southern and western parts of Cheyenne, DeBoer recommended that private property owners plant fruit trees, especially apples and sour cherries, and proposed that the municipality invest in what we would now call a greenbelt, to surround the entire city. It remains unclear how much of DeBoer's plan was carried out before the advent of World War I, which caused his plan to be shelved.[21]

While Cheyenne could rightfully claim the title "City of Trees," Denver emerged as the preeminent example of the integration of ornamental horticulture with overall community improvement. Denver's horticultural preeminence occurred not by chance but through the deliberate action of its leading citizens. They understood that for Denver to become a metropolis, the city needed to be physically attractive; it needed to become "easternized." The critical preliminary step had been completion of the rail line connecting Denver to the transcontinental line at Cheyenne in June 1870 and the line from Kansas City to Denver in August that same year, positioning Denver as the commercial hub of the plains and the mountains. In 1870, twelve years after its founding, Denver's population was 4,759. By 1880 it had reached 35,629, and by 1900 it had reached 133,859.

As noted earlier, the first market gardens in and around the city primarily served the mining communities in the neighboring foothills and mountains. Soon, however, Denver itself became the major consumer of local produce. In pressing the case for construction of a central wholesale marketplace, *Field and Farm* reported that by 1887, local vegetables, fruits, and meats were being shipped to Denver through a network of railroads: the Denver, South Park, and Pacific operated an early-morning route from the southwest; the Denver, Utah, and Pacific ran a train in from Lyons and Longmont; and the Union Pacific brought produce from the Greeley area.[22]

Denargo produce market, Denver, ca. 1925. Courtesy, Rocky Mountain Seed Company, Denver.

As downtown Denver began to look more like a city, with broad streets and blocks of buildings rising to eight and ten stories, elegant new neighborhoods were established in undeveloped parts of the city. In the Capitol Hill area east and southeast of the state capitol, for example, wealthy merchants, ranchers, and silver barons built large residences surrounded by ornamental gardens, broad lawns, and shade trees. "The chief pride of Denver is its homes," reported *New England Magazine* in 1892, "which rank with those of the leading eastern cities, and in point of elegance and comfort are unsurpassed by those of any city of its size in America."[23]

Such praise for Denver was not gratuitous. Recognizing that attractive surroundings drew additional wealth and new residents, Denver civic leaders made sure the eastern press reported on the city's progress. And report it did. When one sees the boulevards and well-landscaped residences, *Harper's* marveled in 1888, one tends to forget that the site of this city was barren waste just twenty-five years ago. Every tree one sees was planted by someone. Nature gave Denver the mountains, bright skies, and limitless plains; beyond that, all else is manmade. And in 1893 *Harper's* exulted: "Denver is a beautiful city—a parlor city with cabinet finish—and it is so new that it looks as if it had been made to order, and was just ready for delivery. . . . The first things that impress you in the city are the neatness and width of the streets, and the number of young trees that ornament them most invitingly."[24]

Despite the publicity, some citizens believed their leaders were not doing enough to beautify the city. Among the critics were Martha A.B. Conine and Mrs. Jasper D. Ward, founders of the Denver Women's Club in 1894. Within four years the club boasted 1,000 members, and its offshoot, the Civic Improvement Society, successfully persuaded city officials to maintain public lawns free of weeds, to water trees, and to place trash cans throughout the city. Because the municipality did not provide adequate street cleaning service, the ladies persuaded private merchants along Sixteenth Street between Arapahoe and Curtis to hire their own street cleaner.

By 1900 Denver had over 800 miles of streets, but only 24 miles in the downtown business district and on Capitol Hill were paved. And like other cities, Denver continued to have its share of misery and dirt, causing the women's organizations to lobby property owners to allow tenants to plant vegetable gardens on vacant lots. The aim was both aesthetic and humanitarian: to make the city more attractive by removing trash and eliminating weeds and, at the same time, to provide useful work and healthful food for the poor. Cultivation of these urban plots, known as Pingree Gardens after the Detroit mayor who had established vacant-lot gardens for the unemployed during the 1894 depression, apparently did not last long. They did, however, provide precedent for Victory Gardens and subsequent community gardens.

From specific beautification activities, the Civic Improvement Society expanded its horizons by establishing an umbrella organization, the Denver Municipal Art League, "to secure united action in the promotion, erection and protection of public works of art and of artistic municipal improvements." Somewhat misnamed, the league consisted of representatives from the Civic Improvement Society, the Women's Club, the Colorado Chapter of the American Institute of Architects, the Artist's Club, and five other organizations. To secure credibility, the league included, as ex-officio members, representatives of the mayor, Park Commission, Board of Public Works, and State Capitol Commission.[25]

While Denver civic leaders appear in most instances to have taken better care of themselves than they did of their fellow citizens, no question remains about their commitment to public parks. That commitment began during the mayoralty of Richard Sopris (1878–1881) and continued through his tenure as commissioner of public parks (1881–1901), when the City of Denver purchased several hundred acres, most notably for City Park, Congress Park, and Washington Park.[26]

To accommodate the increasing demand for landscape irrigation, the privately owned Denver Union Water Company undertook construction of Cheesman Dam and Lake on the upper South Platte River (1893–1905), which, together with existing artesian wells, would supply the city with enough water into the 1930s.[27] By purchasing Denver Union in 1918, the City of Denver inaugurated one of the most progressive municipal water management programs in the country. It was the Denver Water Board that later trademarked and sponsored the development of xeriscaping, beginning in the 1980s.

Firmly establishing Denver as a city beautiful by skillfully waging a campaign against both poverty and ugliness was the work of its remarkable mayor, Robert Walter Speer (1855–1918). Emigrating from Pennsylvania in 1880 in search of a climate more favorable to his tubercular condition, Speer began his Denver career as a carpet salesman for Daniels and Fisher. He moved to an office position because carpet dust and lint irritated his lungs, then left the department store to become a realtor. Four years after arriving in Denver, Speer won election to city clerk and thereby began his career in city government. Because Denver did not yet have a municipal charter, Governor John N. Routt appointed Speer to the Denver fire and police board in 1891, Governor Alva Adams made him police commissioner in 1897, and Governor James B. Orman appointed him president of the Board of Public Works in 1901. As a result of holding these multiple positions, Speer came to control appointments to over half of all city jobs.

Next, Speer played a major role in securing an amendment to the Colorado Constitution that allowed for consolidation of the City and County of Denver. In March 1904 the citizens of Denver adopted a city charter, and in June they elected Speer the city's first "home-rule" mayor. He was reelected in 1908 and again in 1916.[28]

Six months after his first election, at a conference sponsored by the Artist's Club, the mayor outlined the benefits of beautification for the overall betterment of the city. Ever aware of Denver's enviable location, he began by extolling what nature had given the city—pure air, bright sunshine, blue sky, and a 200-mile-wide view of the Rockies—and lamented that nature's gift had been taken too much for granted. He also found worthy of praise all that Denver citizens had accomplished in making their city livable, especially considering its relative youth. But not nearly enough had been done for beautification. To be sure, the city first had to provide for basic necessities, which

Marion Street Parkway designed by Saco Rienk DeBoer, landscape architect, Denver, ca. 1907–1913. Courtesy, Denver Public Library.

we now call infrastructure; nevertheless, even those could have been built more attractively. Take the case of Cherry Creek, which flows through the city from southeast to northwest: it required some concrete walling and artificial channeling for flood prevention, but, by constructing a shady parkway along its banks, neighborhood property values would increase and what was now ugly would become attractive. "Ugliness is not necessary to business," Speer concluded after speaking in favor of street improvements, weed control, dust abatement, and attractive public lighting. For the business district in particular, he noted that electric lines needed to be placed underground and that signs and billboards needed to be controlled, although not banned, by ordinance.[29]

It is no exaggeration to say that Denver's overall appearance changed completely during Speer's mayoralty: parks were greatly expanded, a boulevard and parkway system was constructed, and the number of shade trees, on both public and private property, increased by at least 25 percent. In 1904,

Dailey (Playground) Park designed by Saco Rienk DeBoer, Denver, ca. 1910–1931. Courtesy, Denver Public Library.

city parks covered 573 acres; thanks in part to successful bond campaigns, that figure had increased to 1,184 acres by 1912. As an indication of his approach, the mayor ordered all "keep off the grass" signs removed from city parks because he wanted citizens to use the parks; he also increased maintenance staff so trash could be collected every evening and made sure the numerous park ponds were stocked for fishing at no cost to the citizenry.

Although some of the city's older and wealthier districts were well planted with trees, newer and more modest districts were not. Thus, through the municipality, the mayor inaugurated the free distribution of trees, with accompanying instructions for planting and maintenance. To qualify, a citizen needed to present a written form with an alderman's or the mayor's signature, for which the citizen received one elm and two maples or one maple and two elms (varieties unspecified). On Arbor Day 1905, the municipality distributed 4,992 saplings, and in 1912 it distributed 18,000—a total of 111,000 trees over seven years, with an estimated survival rate of 75 percent. To pay for

Family outing at a Denver park, ca. 1910. Courtesy, Colorado State Historical Society.

the trees, the municipality annually received $5,000 from the Denver City Tramway Company, part of a $60,000 annual gift designated for beautification that Speer had imposed as a condition of an exclusive franchise with the city.[30]

The mayor was unshakeable in his love for trees. During his third term, according to an often-repeated story, the city forester briefed him on the progress of beautification along Cherry Creek. Carolina poplars had been planted between slower-growing elms to provide quicker shade. According to the forester, the time had come to remove the shorter-lived poplars. To which Speer reportedly said: "Cut down those trees? No, sir! Not while I am mayor." "But, Mr. Speer," the forester explained, "it is for the good of the elms I am speaking. They need room to grow." The mayor responded, "I don't care what you are speaking for. They are not going to be cut down. And that ends it, plunk!"[31]

In a speech titled "Give While You Live" before the Denver Civic and Commercial Association in 1916, Speer summarized his career-long, mostly successful campaign to persuade private citizens to contribute monies to public

Backyard, 3020 Lafayette, Denver, ca. 1905–1910. Courtesy, Denver Public Library.

beautification. Appealing to the desire we all have to be well remembered, he noted that "one of the most neglected ways in which people can make themselves bigger and better is by helping to make the city in which they live more attractive." He added: "Ugly things do not please. It is so much easier to love a thing of beauty—and this applies to cities as well as to persons and things. Fountains, statues, artistic lights, music, playgrounds, etc., make people love the place in which they live."[32] By the time of his death in 1918, Speer had attracted over $500,000 in private gifts for beautification—not much considering the wealth present, but a major step toward creating a philanthropic tradition in what was still a young city.

Despite his big-city boss reputation—he supposedly said "I am a boss, and I want to be a good one"—Speer has been described accurately as "a civic idealist who argued that one way to improve humanity was to offer it an uplifting urban environment."[33] As an idealist, Speer strongly endorsed the City Beautiful movement, which stood for beautiful yet functional urban design as a means of making cities more livable. Introduced from France by way of the

144   Forging New Paths in Ornamental Horticulture

1893 Columbian Exposition in Chicago, its principles were first incorporated in the 1901 plan for revitalization of Washington, D.C., based on an original, although largely unrealized, plan of Pierre L'Enfant.

Mayor Speer knew about the Washington plan and contracted with individual members of the commission charged with its execution, among them Frederick L. Olmsted Jr., the leading American landscape architect of the period. In addition, to expedite completion of Denver's boulevard and parkway system, Speer contracted with another City Beautiful advocate, George E. Kessler, a German-born city planner and landscape architect who had distinguished himself as designer of the parkway system for Kansas City, Missouri.[34]

Given the centrality of ornamental horticulture in making the city more beautiful, it is not surprising to discover that during Mayor Speer's third term, a group of interested citizens established the Denver Society of Ornamental Horticulture "to promote beauty in public and private grounds." Its first president was Adam Kohankie, superintendent of Washington Park; its membership, although obscure, appeared limited to parks employees and landed property owners: "there seems to be considerable . . . awakening among our people who have the grounds and the opportunity to ornament the landscape," observed *Field and Farm*, "and why not for it certainly has much to do with the pursuit of happiness in our every day existence. It is time to set aside our [primitive] ways and adopt the ornate things of modern civilization."[35] We know that the society sponsored its second annual flower show at City Auditorium in October 1917 and that the outstanding exhibit consisted of a vast display of chrysanthemums—genera of the composite family and popular subjects for research and cultivation at the Cheyenne Horticultural Field Station beginning in the 1930s.[36]

Long before that, though, the City of Denver under Mayor Speer had established its own nursery to supply the needs of municipal parks, boulevards, and other public places, as well as for distributions to private citizens. At one time, about 60,000 trees and shrubs were under cultivation, including unspecified varieties of the elm, poplar (aspen, cottonwood), maple, sycamore, birch, pine, juniper, plum (wild cherry), and olive (ash), as well as more than twenty varieties of lilacs (*Syringa*).[37]

Increasing us of trees, shrubs, and other ornamental plants naturally created opportunities for local nurseries and seed houses. We have noted that, since the earliest horticultural activities along the Front Range, orchardists in particular had pushed for the development of local stock. Nathan Meeker himself had started a small nursery, but it failed for unknown reasons.[38]

So far as is known, Arthur E. Gipson (1848–1937), who gave his occupation as "lawyer" when he emigrated from Wisconsin to the Union Colony in 1871, established the first viable nursery in Colorado while employed as Greeley postmaster (ca. 1876–1884). He had begun as a fruit grower, presumably obtaining by rail his initial orchard stock—root grafts of pears and at least nine varieties of apples including the then-popular Wealthy, Ben Davis, Red Astrakhan, and Duchess of Oldenburg cultivars. He also successfully cultivated several unspecified varieties of strawberries, raspberries, blackberries, currants, gooseberries, and grapes. In time, adjacent to his "Garden Side Fruit Farm," Gipson planted twenty-five acres to shade and ornamental trees, primarily white ash (*Fraxinus americana* L.), which became part of the Greeley Nurseries. As a founding member of the Colorado State Horticultural Society, Gipson was a tireless critic of unscrupulous "tree peddlers" and an early advocate for publicly supported horticultural experiment stations.[39]

In 1886, Gipson purchased forty acres of orchard land in Wheat Ridge, northwest of Denver, where he established the Colorado Nursery Company. Three years later he sold the Greeley Nurseries, apparently because he thought there was not enough local business, and eventually disposed of all his Colorado properties; in 1891 he moved to Idaho, where he became editor and publisher of a horticultural magazine.

Although the connection is obscure, W. W. Wilmore of Wheat Ridge also started in 1886, with the first nursery along the Front Range to concentrate on ornamental plants. Wilmore committed one-third of an acre exclusively to forty varieties of dahlias (composites), which he cultivated for their variegated single and double flowers; he soon discovered that growing ornamentals was far more lucrative than potatoes on the same plot of land.[40]

While it is unclear where Wilmore purchased his first ornamental seeds, we do know that by the mid-1880s there existed at least one Denver supplier of vegetable seeds and grains. Barteldes and Co., also known as Colorado Seed House, presumably purchased its stock from established seed houses such as the D. M Ferry Company of Detroit (est. 1856) and the W. Atlee Burpee Company of Lancaster, Pennsylvania (est. 1876). Beginning in 1883, the Board of Agriculture authorized Colorado Agricultural College to distribute seeds not used for research, without cost, to farmers and gardeners provided the recipients report their results back to the college.

This matter of free seed distribution had long been of interest to the federal government and, before that, to the founding fathers. In 1839 Congress had approved, and President Martin Van Buren signed, the first appropriation to the Patent Office (precursor to the Department of Agriculture) in support of the free distribution of seeds. Initially, about 30,000 seed packets were distributed annually through congressional offices and agricultural societies. In return for free seeds, farmers were expected to, but rarely did, report their results back to the federal agency. Both J. Sterling Morton and James Wilson, as secretaries of agriculture, tried unsuccessfully to stop the practice of free distribution, which, by the time of their tenure (1893–1913), had turned into petty political graft. Congress was embarrassed into ending it, but not before 1923.[41]

Opposition to "the great free seed farce" by *Field and Farm*, among others locally, can be explained by an interest in protecting and promoting a rapidly growing local seed industry. As early as the 1880s, J. W. Eastwood and J. E. Gauger, farmers near Rocky Ford in the Arkansas River valley, created the "Netted Gem" cultivar of cantaloupe (*Cucumis melo* L.), renamed "Rocky Ford" in 1897. Over the next two decades, farmers in southeastern Colorado expanded their operations to specifically grow seeds from melons to squashes, beans, peas, onion sets, seed potatoes, and other marketable vegetables. The combination of favorable growing climate (cool nights and hot days), aridity to mitigate disease, use of alfalfa as a soil restorative, and irrigation water helped make Colorado's Arkansas River valley the hub of seed production on the High Plains. Among the large seed houses investing early in the valley was the Ferry Company, pioneer in packaging garden seeds in small envelopes for retail sales.[42]

Development of the early seed industry occurred without direct participation by the Colorado Agricultural Experiment Station; its early distinction rested on research in the chemistry of soils. The station's first horticulturist arrived in Fort Collins in 1893: James Cassidy (1847–1889), an Englishman, had worked as a florist at Kew Gardens before immigrating to the United States and earning a degree at Michigan Agricultural College. While his principal research, in line with the college's strength, concerned the chemical analysis of Colorado grasses, he did experiment with insects and insecticides and laid out a plan for broader horticultural experiments on the "fertilization and cross-fertilization of useful plants, with a view to the origination of improved varieties," and "observation and study of the leafage, growth, hardiness and availability of species and varieties of fruit and forest tree growth in Colorado."[43]

Fruit experiments by both Cassidy and his successor, Charles S. Crandall (1852–1929), built on the work by the pioneer orchardists who had settled in the Fort Collins area in the 1870s. Charles Pennock, for example, had begun to cross native with cultivated varieties of fruit trees to create varieties

Tree pruning demonstration, La Porte, Colorado, 1919. Courtesy, Colorado State University Archives and Special Collections, Cooperative Extension Collection.

that were both hardy and flavorful. Crandall did some general work on plums and strawberries and advocated legislative action against the Russian-thistle, contemporary with Bessey in Nebraska and Nelson in Wyoming. Indeed, Bessey's star student, Per Axel Rydberg, prepared and published the first flora of Colorado after Crandall, who had done some collecting, left in 1899 to pursue plant breeding at the Illinois Agricultural Experiment Station.[44]

At the turn of the twentieth century, plant breeding consisted of crossing or transferring pollen from one flower to another of the same or different varieties within a species and hybridizing or transferring pollen between flowers of different species.[45] The preeminent breeder for the High Plains, although he lived and worked in eastern South Dakota, was Niels E. Hansen. His work proved fundamental to the further development of both urban and rural horticulture.

Notes

1. "The Sociological Status of Our Farmers," *Field and Farm* 22 (December 29, 1906): 7.
2. *Field and Farm* 1, no. 15 (April 10, 1886): 9.
3. Ibid., 1, no. 10 (March 6, 1886): 6; 1, no. 18 (May 1, 1886): 7.
4. Colorado State Horticultural Society Annual Report (1890): 568.
5. *Field and Farm* 11, no. 538 (April 25, 1896): 3.

6. Helen L. Grenfell, "The Public Schools and Horticulture," *Colorado Board of Horticulture Annual Report* 13 (1901): 178–179.

7. Edwin I. Grenfell, *A Brief Sketch of the Life and Works of Helen Thatcher Loring Grenfell* (Denver: Smith-Brooks, 1939), 12–14.

8. Grenfell, "Public Schools," 180–184.

9. Mrs. A. E. Gipson, "Children in Horticulture," *Colorado State Horticultural Society Annual Report* (1887): 130–133.

10. Address by Helen L. Grenfell, *Colorado State Horticultural Society Annual Report* 16 (1904): 77–79.

11. Roger L. Williams, *"A Region of Astonishing Beauty": The Botanical Exploration of the Rocky Mountains* (Lanham, Md.: Roberts Rinehart, 2003), 149–154.

12. *Field and Farm* 17, no. 848 (March 29, 1902): 3.

13. C. S. Harrison, "The Home Beautiful," *Field and Farm* 28, no. 1428 (June 21, 1913): 1.

14. *Field and Farm* 17, no. 868 (August 16, 1902): 4; Lyle A. Dorsett and Michael McCarthy, *The Queen City: A History of Denver*, 2nd ed. (Boulder: Pruett, 1986), 83. On restoring the county fair as educational institution, see Liberty Hyde Bailey, *The Country-Life Movement in the United States* (1911; repr. New York: Macmillan, 1920), 167–169.

15. *Field and Farm* 13, no. 635 (March 5, 1898): 8.

16. "Greeley, 1889–90," *Greeley Tribune Extra* (Fall 1889): 1.

17. John Leavy obituary reprinted from the *Greeley Tribune*, Greeley Municipal Archives, Greeley.

18. *Field and Farm* 24, no. 1218 (May 8, 1909): 3.

19. Mrs. Ed W. Sayre, "Organization of First Womans [sic] Club in the Valley," in Asa B. Wood, ed., *Pioneer Tales of the North Platte Valley and the Nebraska Panhandle* (Gering, Neb.: Courier, 1938), 45; see also Pauline Anderson, "The Founding and Early Years of Eaton, Colorado," *Colorado Magazine* 18, no. 2 (March 1941): 55–57.

20. "Report of Citizens Tree Committee," excerpt from Cheyenne City Council proceedings, August 5, 1902, Wyoming State Archives, Cheyenne.

21. S. R. DeBoer, "City Plan. Cheyenne, Wyoming. Street Tree Planting System," typescript [1911?], Wyoming State Archives, Cheyenne.

22. *Field and Farm* 3, no. 12 (March 19, 1887): 3.

23. Dorsett and McCarthy, *Queen City*, 87–88 (quote on p. 87).

24. Edwards Roberts, "The City of Denver," *Harper's New Monthly Magazine* 76, no. 456 (May 1888): 945; Julian Ralph, "Colorado and Its Capital," *Harper's New Monthly Magazine* 86, no. 516 (May 1893): 936.

25. William H. Wilson, *The City Beautiful Movement* (Baltimore: Johns Hopkins University Press, 1989), 172–173.

26. Dorsett and McCarthy, *Queen City*, 110–111.

27. Carl Abbott, Stephen J. Leonard, and David McComb, *Colorado: A History of the Centennial State*, rev. ed. (Boulder: Colorado Associated University Press, 1982), 241.

28. Edgar C. MacMechen, *Robert W. Speer, a City Builder* (Denver: Smith-Brooks, 1919), 13–14.

29. Wilson, *City Beautiful*, 178–180.

30. Ibid., 183; MacMechen, *Speer*, 15–16, 18.

31. MacMechen, *Speer*, 56.

32. Reprinted in ibid., 75–77.

33. Wilson, *City Beautiful*, 170.

34. Ibid., 181; Abbot, Leonard, and McComb, *Colorado*, 248.

35. *Field and Farm* 31, no. 1569 (February 26, 1916): 3.

36. Ibid., 32, no. 1655 (October 27, 1917): 3.

37. Ibid., 26, no. 1315 (April 15, 1911): 3, and 26, no. 1326 (July 1, 1911): 3.

38. David Boyd, *A History of Greeley and the Union Colony of Colorado* (Greeley: Greeley Tribune Press, 1890), 158.

39. Biographical note from "Records of Meeker Memorial Museum," n.d., Greeley Municipal Archives; "Farming in Colorado under the System of Irrigation. Also a Short Review of the Climate and Resources of Weld County and the City of Greeley" (Greeley: Sun Publishing, 1887), 22.

40. *Field and Farm* 3, no. 9 (February 26, 1887): 4; 14 (October 1, 1887): 7.

41. Gladys L. Baker, Wayne D. Rasmussen, Vivian Wiser, and Jane M. Porter, *Century of Service: The First 100 Years of the United States Department of Agriculture* (Washington, D.C.: USDA, 1963), 5–6, 13, 34, 43, 106.

42. Robert G. Dunbar, "History of [Colorado] Agriculture," in Leroy Hafen, ed., *Colorado and Its People* (New York: Lewis Historical Publishing, 1948) 2: 143–144; Frank J. Annis, "Concerning the Duties of the Secretary of the State Board of Agriculture, and the Distribution of College Seeds and Plants," *Colorado Agricultural Experiment State Bulletin* 3 (December 1887): 1; *Field and Farm* 31, no. 1613 (December 30, 1916): 8.

43. *Field and Farm* 1, no. 10 (March 6, 1886): 3; Joseph A. Ewan, *Biographical Dictionary of Rocky Mountain Naturalists* (Utrecht: Bohn, Scheltema and Holkema, 1981), 38; quote from Colorado Agricultural Experiment Station Annual Report (1888): 7.

44. C. L. Ingersoll, "Experiment Stations, the Relations of, to Horticulture," Colorado State Board of Horticulture Annual Report (1889): 205–207; Ewan, *Biographical Dictionary*, 51; Charles S. Crandall, "Notes on Plum Culture," *Colorado Agricultural Experiment Station Bulletin* 50 (1898), 48 pp.; Crandall, "Strawberries," *Colorado Agricultural Experiment Station Bulletin* 53 (1900), 27 pp.; "The Russian Thistle," *Colorado Agricultural Experiment Station Bulletin* 28 (1894), 12 pp.; Per Axel Rydberg, "Flora of Colorado," *Colorado Agricultural Experiment Station Bulletin* 100 (1906), 448 pp.

45. Wendell Paddock, "Improvement of Plants for Colorado by Breeding," Colorado Board of Horticulture Annual Report (1902): 143–148.

9

Collecting and Creating Hardy Plants

Readers of Aven Nelson's "Horticultural Column" in the December 1912 *Wyoming Farm Bulletin* might have seen the notice about Niels Hansen, secretary of the South Dakota Horticultural Society, as well as professor and horticulturist at the agricultural experiment station in Brookings. "As is well known," Nelson wrote, Hansen "has succeeded in developing some remarkably promising fruits." In particular, for the higher elevations of Wyoming, some of Hansen's creations "work a revolution," combining the size and quality of tender cultivars with the hardiness of native wild plants. Nelson urged interested readers to get in touch with Hansen personally or with Gurney's of Yankton, South Dakota, the nursery and seed house that was handling Hansen's products.[1]

At one time or another during the first half of the twentieth century, every agricultural experiment station and commercial nursery on the High Plains carried some variety of fruit, vegetable, ornamental, tree, or shrub either introduced from foreign countries by Hansen or developed by him at the

South Dakota Agricultural Experiment Station. Indeed, Hansen had gained notoriety, initially because of reports of his adventures as plant explorer in northern Europe and Asia, and then as the result of his successes as self-promoter. It was said that wherever he traveled, at least in the Dakotas, his first stop was the local newspaper office. Popular national magazines such as *Better Homes and Gardens* took up his story as plant breeder under headings such as "The [Luther] Burbank of the Plains," Burbank regarded as America's leading plant breeder.

In all fairness, Hansen was an enormously energetic individual. He developed around 300 varieties of fruits, vegetables, ornamentals, trees, and shrubs; he collected and introduced numerous hardy forage plants and grains—most notably, subspecies of alfalfa (*Medicago sativa* L.), crested wheat (*Agropyron cristatum* L. (Gaertn.)), and smooth brome (*Bromus inermis* Leyss.)—all now naturalized on the High Plains.

Born in Denmark, Niels Ebbesen Hansen (1866–1950) immigrated to the United States at age seven with his family, In 1877 they reached Des Moines, Iowa, where his father, an interior decorator, had secured a long-term commission to paint frescoes in the state capitol. The younger Hansen, meanwhile, came to the attention of John A. Hull, Iowa's secretary of state, who gave him a job as messenger, tutored Hansen in basic subjects, and prepared him to enter college in 1883.[2] Iowa Agricultural College was the nearest well-established college.

Hansen's decision to major in botany and horticulture remains obscure. Comparing the two disciplines, he wrote his father from college that "in Botany we learn to draw and to know the difference between all the many different types of trees, flowers, etc. In Horticulture we learn to handle and grow raspberries, strawberries, currants, grapes especially and many other small fruits."[3] One senses a preference for the immediate, economic applications of science rather than for the discipline of the search for knowledge in itself.

Hansen learned botany from Charles Bessey, who was still at Iowa; as a student assistant, he worked in Bessey's "experimental garden." He took a course in "Agriculture" from Seaman Knapp and worked on the college farm under Knapp's direction. His major professor was the horticulturist Joseph L. Budd (1835–1904), best known for introducing Russian varieties of fruits and shrubs to the Great Plains. Budd was part of that long tradition of Americans searching the world for useful plants, stretching back to 1787 when Thomas Jefferson collected and sent home from Italy the seeds of upland rice. A self-taught expert on fruit culture, Budd had started out as a nurseryman in upstate New York, where he befriended the pomologist Charles Downing (brother of the pioneer landscape architect Andrew Jackson Downing). In 1885 Budd inherited Downing's library and notebooks, still housed at Iowa State University.[4] Hansen, in other words, had at his disposal numerous horticultural classics including Alphonse de Candolle's *Origin of Cultivated Plants*, of which more follows.

After earning the baccalaureate in 1887, Hansen took employment in commercial nurseries. In 1892, Budd invited him back to complete a master's program and provided him with an assistantship that required both research at the experiment station and service off campus through lectures at Farmers' Institutes.[5] As a graduate student, in 1894 Hansen had the extraordinary experience of a four-month "horticultural tour" of northern Europe and Asia. Later, he attributed to this trip the formulation of what became the principle underlying his plant-breeding work: to make certain warm-climate plants hardy enough to thrive on the northern Great Plains, they must be crossed or hybridized with plants that are naturally hardy.

By hardiness, Hansen meant more than resistance to cold; he also meant resistance to heat, drought, wind, and the other extremes found on the northern Great Plains and, by extension, the High Plains. And while he would never be in the position to isolate the specific property that produces hardiness—electron microscopes had yet to be invented—he knew that hardiness could be transmitted through plant breeding.[6]

Finding a hardy winter apple was perhaps the most urgent fruit problem of the day, so it is not surprising that Budd directed Hansen to prepare a master's thesis on "northwestern apples." While the thesis, completed in 1895, is apparently lost, two subsequent publications, very similar to each other, undoubtedly stem from that work. Hansen's aim was to describe and make drawings of every cultivated variety in Iowa, Minnesota, and South Dakota. By "variety," Hansen meant a "sport" or plant that exhibits some striking variation, such as cold hardiness, from its parent stock, not "variety" in the way a botanist might use the term "subspecies."

After studying the European system of classifying apples into what Hansen called "families based on natural affinities," which he found too complicated for general use, he developed an artificial key based on both internal and external characteristics of the fruit. Until a truly hardy apple was bred for the northern Great Plains, Hansen argued, orchardists needed to choose carefully from the varieties available. Attempting to classify the seemingly hundreds of apple varieties posed inherent difficulties. Most notably, varieties of apples do not grow "true to seed," that is, ten seeds saved out of a single apple may produce ten varieties widely differing in color, size, quality, and date of maturity. Hansen had observed that such varieties were generally inferior and that, if a variety worthy of propagation appeared, it would be far easier for the gardener or orchardist to reproduce that chosen variety by grafting, budding, or root sprouts rather than trying to stabilize or "fix" it through a long process of selection.[7]

By the time Hansen wrote his thesis, horticulturists generally recognized that all standard varieties of apples had originated from the apple native to Asiatic Russia (today, Kazakhstan) and that because of its broad geographic range, the apple clearly had produced, over a long period of time, variations in hardiness according to locality. Recognizing that apple growers on the northern Great Plains could not wait for the "slow process of acclimation by nature" to produce a hardy variation for their climate, Hansen recommended that only the hardiest known varieties, mostly Russian, be planted—at least until a suitable variety could be bred.

To justify his assertion that horticulturists could not introduce hardiness by selection alone, Hansen referred to a passage in Alphonse de Candolle's *Origin of Cultivated Plants*, translated from French in 1886. De Candolle had observed that "the northern limits of wild species . . . have not changed within historic times although the seeds are carried frequently and continually to the north of each limit. Periods of more than four or five thousand years . . . are needed apparently to produce a modification in a plant which will allow it to support a greater degree of cold."[8]

De Candolle's observation, which Hansen cited frequently, provided the fundamental intellectual challenge for his plant-breeding career. To accelerate nature's work, Hansen sought to combine the winter hardiness of the Siberian crab apple, *Pyrus baccata* L. → *Malus baccata* (L.) Borkh.; the hardiness and storage quality of the prairie crab apple, *Pyrus ioensis* (A. Wood) L.H. Bailey → *Malus ioensis* (A. Wood) Britt.; and the quality fruit of the standard European cultivated apple, *Pyrus pumila* (Mill.) C. Koch, not J. Neumann ex. Tausch → *Malus pumila* Mill. While Hansen ultimately failed in his effort to breed the ideal apple for the northern Great Plains, in 1918 he did introduce South Dakota No. 2, a seedling of prairie crab apple top grafted on the Russian Duchess of Oldenburg, the earliest maturing apple then known, which he hoped would become popular in the region.[9] That apple ("Anoka"), which has greenish-yellow fruit streaked with red, is no longer sold because of its relatively short life, susceptibility to fire blight, and poor storage quality.

Underlying Hansen's breeding work was the assumption that all species are in the process of continual change. In that way, Hansen followed Charles Darwin (1809–1882), who had rejected the long-accepted notion that species are permanently fixed. Darwin had popularized those mechanisms by which he thought such change could be explained: natural selection, sexual selection, and the inheritance of acquired characteristics. By natural selection, Darwin meant that those individuals within a given species with chance variations favorable to survival tended to survive and reproduce, while those individuals without such chance variations tended not to survive; further, repeated favorable variations through many generations would eventually result in a new species. Consistent with this view, Hansen argued that breeding for hardiness was a way to dramatically accelerate the process of natural selection.[10]

In *Variation of Animals and Plants under Domestication* (1868), which Hansen had read, Darwin hypothesized—incorrectly, it turns out—that characteristics are inherited through a process he called pangenesis, whereby bodily (somatic) cells of parent organisms release tiny particles, or "gemules," that come together in reproductive (germ) cells and thereby create entirely new organisms. Later, August Weismann (1834–1914), who accepted the Darwinian premise of evolutionary change, nevertheless rejected the possibility that an organism could inherit acquired characteristics. Weismann argued that germ cells are created separately and not from somatic cells. Since acquired characteristics are variations in somatic cells and somatic cells die with the individual organism, acquired characteristics cannot be inherited.[11] Hereditary characteristics are transferred only through germ cells; thus, another mechanism is needed to account for the existence of variation, whether by natural selection or artificial breeding.

Hansen referred to that other mechanism as Mendel's Law. At the First Congress of Genetics in New York City in 1902, he heard geneticist William Bateson of Cambridge University deliver one of the first lectures on Mendel ever given in the United States; at the Louisiana Purchase Exposition (World's Fair) in St. Louis in 1904, Hansen listened to Hugo DeVries of Amsterdam University, who devoted much of his scholarship to rediscovering and publicizing Mendel's work.[12] Gregor Mendel (1822–1884) was an Augustinian monk, abbot of Brüun in Austria, and amateur naturalist. In the garden of his monastery, he had experimented with crossbreeding different varieties of the common pea (*Pisum sativum* L.). Unlike Darwin, who concentrated on mechanisms of change, Mendel sought out mechanisms to explain stability in species. By controlling plant fertilization—carefully removing pollen from the anthers of the stamen and manually introducing it to stigma of the ovary—Mendel discovered that he could breed for certain characteristics and that those characteristics would appear in subsequent generations in proportions he could predict mathematically.

In studying the principal characteristics of peas, Mendel noticed that each characteristic occurred in two contrasting forms: seeds were either smooth or wrinkled, flowers were either white or pink, pods were either green or yellow, and so on. Furthermore, in cross-pollinating, for example, one parent plant (P) with green pods and one parent (P) with yellow pods, he found that all plants of the first generation (F_1) had either all green or all yellow pods. But when he crossbred all plants of the first generation (F_1) among themselves, he found that some second-generation plants (F_2) had green pods and others had yellow pods. In other words, he noticed that second-generation plants (F_2) exhibited characteristics present in the grandparental generation (P) but not in the parental generation (F_1). Mendel called the characteristics appearing in the first generation dominant and those that remained latent in the first generation recessive.

Besides being able to calculate mathematically the frequency of dominant over recessive characteristics through multiple generations, Mendel deduced that contrasting characteristics such as green or yellow pods must exist as separate and distinct entities within the cell and that they cannot be blended or fused with each other. He had indeed found stability within evolutionary change. Hansen took Mendel's experiments to mean that if we make enough crosses through enough generations, we will get an individual that has any desirable combination of characteristics we wish.[13]

Just as Mendel's experiments depended on painstaking hand-pollinating over many generations, so too Hansen's breeding for hardiness would involve massive amounts of what horticulturists once called "emasculating." And Darwinian principles contributed the rationale for Hansen's expeditions in search of plants that had hardy characteristics to crossbreed with those that did not. Between 1897 and 1924, Hansen made six trips throughout northern Europe and Asia, not to mention his numerous shorter trips into the Dakotas, adjoining states, and Canada.

Hansen undoubtedly "caught" the Darwinian spirit at Iowa. Through his experiment station work, he also made the acquaintance of its director, James "Tama Jim" Wilson (1835–1920). Upon becoming secretary of agriculture in 1897, Wilson's first directive was to send Hansen, then professor of horticulture at the South Dakota Agricultural College, to Russia in search of hardy, cold-resistant cereals and fruits. In point of fact, Hansen succeeded in collecting a then-unknown variety of alfalfa, which he named "Cossack"; importing scions of the Siberian or "Dolgo" crab apple, which he would use extensively for breeding with standard apples; and importing vast quantities of crested wheat, which remains to this day a much-used reclamation grass on the High Plains.[14]

Hansen reported on his travels directly to Secretary Wilson rather than to David Fairchild, head of the Section of Seed and Plant Introduction (created shortly after Hansen's first commission); that violation of bureaucratic protocol sowed the seeds of Hansen's sometimes turbulent relationship with the United States Department of Agriculture (USDA). He was an omnivorous collector, seemingly unconcerned with preliminary classification in the field; so when his collections arrived at the USDA in Washington, section employees found themselves overwhelmed not only with having to classify contents but also with locating parties willing to field test the imported varieties. The fact that Hansen traveled on a reimbursement basis, without the usual budgetary constraints, provided further grounds for antagonism—although apparently not among the Wilson family. In 1902 Hansen's Iowa classmate, James Wilbur Wilson, who had served as private secretary to his father in Washington, became director of the South Dakota Agricultural Experiment Station, a position he held until 1938.[15]

Although Hansen's adventures as explorer in regions of the world generally unknown to Americans are not part of our story, the newspaper publicity about his trips established him as a folk hero in South Dakota. In addition, at the height of his plant-breeding career, Hansen paid a visit—much publicized in South Dakota—to Luther Burbank in Santa Rosa, California.[16] All these efforts enhanced his image in the South Dakota public eye and ensured continual funding for travel and staff out of the state treasury, even after the USDA ended its direct support in 1913 and after Hansen had accepted consulting contracts from the government of the USSR under Stalin in 1934.

By Hansen's own count, he introduced over 300 new varieties to the northern Great Plains between 1895 and 1927, either through importation to this country or through breeding at the South Dakota Agricultural Experiment Station. Many of those varieties proved suitable to the High Plains. Among them, Hansen seemed to take the greatest pride in successfully breeding the hardiness of the native western sand cherry with the eating quality of cultivated Japanese plums to create a cherry suitable for home use on the High Plains. When Hansen informed his former professor, Charles Bessey, about his new hybrids, Bessey expressed great satisfaction "that for a time you were a student of mine, for work of this kind is really great work."[17]

Bessey undoubtedly took special pleasure in the fact that his former student was working on a plant named in his honor. The western sand cherry (*Prunus pumila* L. var. *besseyi* (Bailey) Gl.) is a compact ground or dwarf cherry, much branched, rarely more than three feet high, with dark purple fruit about the size of the commercial cherry. It can be found locally, common

Western sand cherry, Rocky Mountain Herbarium, University of Wyoming, Laramie. Courtesy, Ronald K. Hansen.

in sandy and rocky soil on the High Plains of Nebraska, Wyoming, Colorado, and Kansas. In 1878, Charles E. Pennock reported finding the plant along the Cache la Poudre River upstream from Fort Collins near Bellvue. Calling it "the most valuable fruit I ever saw growing wild," Pennock collected and planted seed of the best fruit in his orchard. In 1892 he introduced his stock to the commercial trade as the "Improved Dwarf Rocky Mountain Cherry." His fellow nurseryman, Arthur E. Gipson of Greeley, observed that the sand cherry produced a fruit "especially valuable for pies and preserves, and is often pleasant to eat from the hand. It is wonderfully productive, and will survive all changes and vicissitudes of the most exacting climate."[18]

In 1889, Charles Bessey introduced the sand cherry to members of the American Pomological Society as "a promising new fruit from the Plains." To his fellow Nebraskans, he later wrote:

> No native fruit appears more promising than this. Even in a wild state it is very prolific, and when fully ripe it is edible in the uncooked state. The astringency which is present in the unripe fruits almost or entirely disappears at maturity. Plants appear to differ a good deal in the amount of astringency, as well as in the size and shape of the cherries which they bear. In many parts of the state the sand cherry has been transplanted to the garden or orchard. Wherever this has been done the results have been encouraging. The plants become larger, and the cherries are larger and more abundant. They root freely from layers [shoots fastened down and partly covered with soil to take root while still attached to the parent plant], and hence are propagated with the greatest ease. My studies of this interesting native cherry, supplemented by the testimony of numerous observers in all parts of the state where it grows, lead me to the conclusion that we have here a fruit which needs only a few years of cultivation and selection to yield us a most valuable addition to our small-fruit gardens.[19]

Niels Hansen began his career-long work on the western sand cherry as a graduate student, assisting Joseph Budd at the Iowa Agricultural Experiment Station. Beginning in 1892, Budd had planted several thousand seeds collected near Valentine, Nebraska; over the next three years, he marked for propagation those seedlings that bore the largest and best-quality fruit. Budd's aim, which Hansen would fulfill, was to use the western sand cherry as rootstock for cultivated plums.

Hansen continued the experiments after he moved to Brookings, using plants already established at the South Dakota Agricultural Experiment Station as well as imports from Iowa. Hansen, too, grew thousands of seedlings to fruit and selected around seventy-five varieties for propagation and further trials. He kept in touch with Bessey, inquiring, for example, about possible additional choice varieties cultivated in Nebraska and other parts of the West.[20]

By selection alone, through fourteen generations (1895–1941), Hansen isolated a choice specimen of western sand cherry, introduced to the commercial market as the Hansen bush-cherry and still available through specialty houses. It produces quality fruit nearly one inch in diameter, twice as large as most sand cherries, with pits about one-fourth their natural size.[21]

In the summer of 1902, Hansen made a wagon trip to the Sioux country of southwestern South Dakota to secure specimens of western sand cherry from the "gumbo" clay lands along the Cheyenne River. In the company of an elderly Sioux, he did find cherries on steep bluffs along the river. He was struck by the Sioux name for sand cherry: "with the wind," apparently meaning that "when picked with the wind the cherries are sour, and if picked against the wind, they are sweet." Hansen found this an ingenious explanation of the unequal quality and size of the cherry's fruit.[22]

By overcoming those inequalities, Hansen expected to create a choice fruit for home use and, most important, to use the western sand cherry as stock to which other species of plums (*Prunus*)—sweet and sour cherries, apricots, peaches, native plums, and others—could be grafted or budded to greatly extend the cultivars' geographic distribution—north to Manitoba and west to the High Plains. Aware that the popular press had given "grafting" a sinister meaning by using it to describe various forms of financial corruption, Hansen reminded readers that his use of the word came from the German *veredlung*, translated as ennobling, improvement, refinement, cultivation: "To cause a wild plum tree, for example, bearing small and inferior fruit, to bear large, choice fruit by the simple operation of budding or grafting is really to 'ennoble' that tree." It is also less of a lottery than raising plums from seed and much quicker than starting them from sprouts or suckers, and it cleverly combines several desirable characteristics in one plant.[23]

Essentially, budding and grafting are the same operation. Both unite the bud of a variety to be propagated with the rootstock from which it develops growth. In grafting, one unites a scion (shoot or twig) with more than one bud to the rootstock. In the case of plums, Hansen preferred the method of crown grafting. For outdoor grafting, which is done in early spring before buds swell and while a plant is still dormant, he inserted a plum scion at or just below the surface of the ground onto sand cherry seedlings established one year in his nursery. For indoor or greenhouse grafting, done during the winter months, he inserted dormant plum scions into sand cherry rootstocks.

Budding occurred only during late summer, when the bark could be easily peeled back. Hansen described the process: "A T-shaped slit is cut in the bark of the stock and the edges lifted with a sharp, thin-bladed knife. Into this opening is inserted the bud, cut with the adhering bark called the shield, which extends about one-half inch below the bud. The shield is cut with a very thin layer of the newly forming wood the full length of the shield. After insertion, the tying is done with raffia." Budding, too, was done as close as possible to the ground; in arid country, the bud was inserted on the north side of the rootstock to provide shade, then covered wholly or in part with soil.[24]

Today, mechanical "T-budding" is the most common grafting used by nurseries, the quickest and most labor-saving method that uses the least amount of scion wood. Hansen knew about machines being developed and patented in his day to automate grafting, but he preferred the manual approach as providing the safest and surest successes.

By grafting the wild plum onto the sand cherry rootstock, Hansen obtained a dwarfed tree that bore fruit early and abundantly but maintained its standard plum size and quality. In some cases, however, the trees grew too quickly for the root systems, so branches drooped under heavy loads of fruit, which was acceptable for small home gardens but not for commercial orchards. As a result, Hansen tried to grow these trees in bush form, with very low stems, and by cutting back the tops during the first years of growth.[25]

In spring 1907, Hansen sent Bessey a handbill on new fruits available through the South Dakota Agricultural Experiment Station and included a note about sending his latest sand cherry–based creation "so you can make a scientific report from the botanist's standpoint on this hybrid." Bessey responded with delight that he had a place for two or three specimens in his home garden and commented later about his ever-increasing interest in his former student's work on breeding cultivars with native fruits.[26]

Hansen's most successful plum hybrids were "Opata" and "Sapa," both introduced in 1908 and grown widely from Oklahoma north to Canada, including the High Plains up to about 6,000 feet. The Opata was a cross between the western sand cherry (female parent) and the gold plum (male parent), itself a hybrid of the Japanese plum (*Prunus triflora* Roxb. → *Prunus salicina* Lindley) created by Luther Burbank. Hansen described the Opata as "a plum tree in habit of vigorous growth and forms fruit buds freely on one-year old shoots in nursery; foliage large and glossy. Fruit, one and three-sixteenths inches in diameter, dark purplish red with blue bloom; weight one-half ounce; flesh green, firm; flavor very pleasant, combining the sprightly acid of the sand cherry with the rich sweetness of the Gold plum. Excellent for eating out of hand. The thin skin can be chewed and eaten, as it is entirely free from acerbity. Pit very small; season extremely early."[27]

Professor Hansen's fruit trees. Courtesy, Archives and Special Collections, South Dakota State University, Brookings.

Like the Opata, the Sapa was a cross between the western sand cherry (female parent) and the sultan plum, also a Burbank hybrid of the Japanese plum. Producing dark purple skin, flesh, and juice, the Sapa bore well on one-year shoots in the nursery and ripened extraordinarily early.[28] In western Nebraska during the 1920s, Jules Sandoz reported successfully growing "Sapa Cherry Plum" and "Opata Cherry Plum"; he acknowledged receiving advice directly from Hansen. Both varieties remain available through specialty nurseries.[29]

Among all the wild fruits of America, the strawberry is arguably the most highly regarded. As early as 1643, the Reverend Roger Williams of Rhode Island wrote that the strawberry "is the wonder of all the fruits growing naturally in those parts; it is of itself excellent, so that one of the chieftest doctors of England was wont to say that God could have made, but never did, a better berry."[30] Anyone living on the High Plains who has ever picked the wild strawberry (*Fragaria virginiana* Duchn.) or the woodland strawberry (*Fragaria vesca* L.) knows that while delicious, the fruit is so small that it would be impractical to cultivate even in the home garden. And the large-fruited cultivars found in supermarkets could not survive winters on the High Plains.

Shortly after his arrival in South Dakota, Hansen began a series of experiments to develop varieties of strawberries and also raspberries that would survive winters without mulch or any other special protection. In 1895 he started by collecting native strawberries from near Brookings north to the Manitoba border, and he secured standard varieties from nurseries. Over a ten-year period, he cross-pollinated more than 8,000 seedlings. This procedure involved carefully removing the pollen from one variety, carrying it to another variety, and—with knife point or small brush—placing the pollen correctly on the stigma, then noting and tagging each cross. Adding to Hansen's difficulty, strawberries often contain staminodes (sterile stamens), which led him to conclude—incorrectly—that the strawberry plant is of two kinds, imperfect and perfect, with the former bearing only pistillate (female) flowers and the latter containing both staminate (male) and pistillate flowers. The reason for the vast numbers of strawberry plants is that cross-pollination, according to Mendelian principles, does not produce plants true to seed—unlike grafting, which, through cuttings (asexual means), produces true clones.[31]

While Hansen never reached the point of creating what he called the "Ideal Farmer's Strawberry," he did introduce two hybrids in 1907. South Dakota No. 1 united pollen from a wild strawberry collected in Manitoba with a "Jessie" cultivar; South Dakota No. 2 united a wild strawberry from northern

Opata plum, Cheyenne Horticultural Field Station. Courtesy, USDA–Agricultural Research Service.

North Dakota with "Glen Mary." The former, about one inch in diameter and of excellent quality, had survived −40°F. temperatures without mulch or snow cover. Similarly, in 1906, Hansen had succeeded in breeding a wild raspberry found in northern North Dakota with the "Shaffer's Colossal" cultivar; he introduced his creation to the commercial trade as the "Sunbeam" raspberry "because it appeared as a sunbeam when the outlook for hardy raspberries was dark."[32]

Hansen's strawberry and raspberry hybrids were grown commercially on the High Plains of southwestern South Dakota by John Stevenston Robertson (1866–1935). A native of Ohio, Robertson had moved with his parents to eastern Nebraska and, from there, to his own homestead near Hot Springs, South Dakota, in 1892. Despite having little formal education, he became a successful dry-land fruit grower, lectured frequently at Farmers' Institutes, and served a long tenure as horticulture editor of *Dakota Farmer*. Each spring he welcomed South Dakota Agricultural College students to experiment with cross-pollination at his orchard; he gladly accepted for trial many of Niels Hansen's creations.[33] Specimens from Robertson's orchard numbered among the earliest accessions of the Cheyenne Horticultural Field Station.

Robertson served as one of a handful of members of the South Dakota Horticultural Society from the southwestern part of the state and succeeded Hansen as society president in 1933. The society had been incorporated in 1890, and Hansen served as its secretary from 1898 until 1929. In that capacity, he also edited the society's annual report, supported by state appropriations since 1904. He used the annual report as a vehicle to promote a horticultural "great awakening" statewide and to reprint his many public utterances on the economic value of homestead and community beautification. As in the neighboring states of Nebraska and Wyoming, the South Dakota Horticultural Society counted a relatively small membership, never more than 300 dues-paying members plus complimentary memberships for local newspaper editors.[34]

Toward the end of his career, Hansen began to use the radio to publicize his horticultural pontifications. In a 1936 broadcast, for example, he advocated the planting of more home gardens within towns for the sake of better health, both physical and mental, and the incorporation of more gardens on school grounds so children could learn more directly about science and its applications. Then, in light of the national emergency at the time, he proposed "An Action Program for Horticulture in Wartime" that called for adding a county horticulturist to the existing team of county agent and home demonstration agent. His idea was for the county horticulturist to serve as the "shock absorber between Science and Practice," working in the garden or the field side by side with the gardener, the farmer, or the farmer's family.[35]

In like manner, Hansen viewed himself as filling a unique niche in the field of horticulture, namely, as plant breeder or, as he preferred to call himself, plant inventor—which private business had neither the expertise nor the finances to support. He left propagation of his invented cultivars to the nurseryman, seedsman, and florist. Specifically, he had developed a warm relationship with Charles W. Gurney after the pioneer nurseryman moved his business from northeastern Nebraska to Yankton, South Dakota, in 1906, and then with Gurney's son, Deloss B., who took over the business in 1913. The "House of Gurney" served as the principal commercial outlet for Hansen's creations as well as his imports until 1922, when his son, Carl Andreas, established the Hansen Nursery on a five-acre plot just east of Brookings. Initial tree and shrub stock came from cuttings made at the South Dakota Agricultural Experiment Station.[36]

While hardy fruits formed the core of the Hansen Nursery, a few vegetables and numerous shrubby plants from abroad were propagated and distributed commercially. Most notable among the imports was the Siberian pea-shrub (*Caragana arborescens* Lam.), of which the elder Hansen was the first to collect the seeds in bulk (350 pounds) during his 1897 trip to Russia. Because of its resistance to drought and cold, caragana remains to this day the most common plant for the outside rows of shelterbelts and for ornamental hedges on the High Plains. Left untrimmed, the caragana grows into a large, somewhat thorny bush, up to fifteen feet in height. In May, the plant is covered with flowers similar to the pea, although the color is yellow; the foliage, which appears relatively early, is a lively green. Its abundant seeds are easy to collect before the pods burst open or dehisce, and they grow easily if sown in late fall.[37]

To the end of Hansen's plant-breeding career, only a few commercial nurseries existed beyond the 100th meridian on the High Plains: at Hot Springs, North Platte, Cheyenne, and, as noted, along Colorado's Front Range from Denver north to Fort Collins and Greeley. As a result, much of Hansen's stock ended up at the region's agricultural experiment stations and at Forest Service and other USDA field stations, including the Cheyenne Horticultural Field Station.[38]

Indeed, sometime during the spring of 1929, after the USDA received orders to construct the Cheyenne facility, E. C. Chilcott, head of dry-land agriculture and former vice director of the South Dakota Agricultural Experiment Station, was dispatched to Brookings to look at Hansen's greenhouse for possible replication and to seek Hansen's counsel on related items. Twenty years later, the name A. C. Hildreth, superintendent of the Cheyenne Field Station, appeared on the special guest list for an assembly at Brookings, presided over by Governor George T. Mickelson, to honor Hansen's contributions to horticulture. Despite his differences with the USDA, Hansen could not dispute that his work was possible only because of federal assistance to the South Dakota Agricultural Experiment Station. Such assistance, both technical and financial, had become widespread over the High Plains during the first three decades of the twentieth century.[39]

Notes

1. Aven Nelson, "New Fruits," *Wyoming Farm Bulletin* 2, no. 6 (December 1912): 244.

2. Helen Hansen Loen, ed., "The Journals of Niels Ebbesen Hansen, 1879–1892" (typescript, 2004), pp. 84–92, carton 1, Niels E. Hansen Papers, South Dakota State University, Brookings (hereafter cited as Hansen Papers).

3. Niels Hansen to Andreas Hansen, Ames, August 17, 1883, tr. Helen Hansen Loen, carton 1, Hansen Papers.

4. Ibid.; Loen, "Journals," 96–109.

5. Loen, "Journals," 142; Niels Hansen to Andreas Hansen, Ames, January 28, 1895, carton 1, Hansen Papers.

6. Niels E. Hansen, "Breeding Hardy Fruits," South Dakota Horticultural Society Annual Report 4 (1907): 190–191; Arnold Nicholson, "Burbank of the Plains," *Country Gentleman* 110, no. 12 (December 1940): 54.

7. Niels E. Hansen, "A Study of Northwestern Apples," *South Dakota Agricultural Experiment Station Bulletin* 76 (June 1902): 3–4; Hansen, "Apples," in Joseph L. Budd and Niels E. Hansen, *American Horticultural Manual, Part II. Systematic Pomology* (New York: John Wiley and Sons, 1903), 26–27.

8. Hansen, "Northwestern Apples," 8; Alphonse de Candolle, *Origin of Cultivated Plants* (1886; repr. New York: Hafner, 1967), 460; see also John E. Miller, "Eminent Horticulturalist: Niels Ebbesen Hansen," in Herbert T. Hoover and Larry J. Zimmerman, eds., *South Dakota Leaders: From Pierre Chouteau, Jr. to Oscar Howe* (Vermillion: University of South Dakota Press, 1989), 273.

9. Niels E. Hansen, "Plant Introductions," *South Dakota Agricultural Experiment Station Bulletin* 224 (1927): 9.

10. Miller, "Eminent Horticulturalist," 273.

11. Niels E. Hansen, "450 Years" (typescript, 1943), carton 2, Hansen Papers; Loren Eiseley, *Darwin's Century* (New York: Doubleday, 1958), 216–219; Roger L. Williams, *Modern Europe 1660–1945* (New York: St. Martin's, 1964), 397.

12. Niels E. Hansen, "Experiments in Plant Heredity," *South Dakota Agricultural Experiment Station Bulletin* 237 (April 1929): 13.

13. Hansen, "Breeding Hardy Fruits," 188–190; Eiseley, *Darwin's Century*, 205–209. For a readable explanation of Mendel's experiments, see William T. Keeton, *Biological Science* (New York: W. W. Norton, 1967), 519–522.

14. Nelson Klose, *America's Crop Heritage: The History of Foreign Introduction by the Federal Government* (Ames: Iowa State College Press, 1950), 109–114.

15. David Fairchild, *The World Was My Garden: Travels of a Plant Explorer* (New York: Charles Scribner's Sons, 1938), 106–108; Isabel S. Cunningham, *Frank N. Meyer, Plant Hunter in Asia* (Ames: Iowa State University Press, 1984), 6, 141, 163; Secretary James Wilson to Niels E. Hansen, Washington, D.C., March 17, 1897, carton 2, Hansen Papers.

16. Niels E. Hansen, "A Visit to Luther Burbank," South Dakota Horticultural Society Annual Report (1906): 112–114.

17. Charles E. Bessey to Niels E. Hansen, Lincoln, February 29, 1908 (microfilm), reel 20, Bessey Papers, University of Nebraska, Lincoln (hereafter cited as Bessey Papers).

18. Liberty Hyde Bailey, *Sketch of the Evolution of Our Native Fruits* (New York: Macmillan, 1898), 233–242 (Pennock quote on p. 242; Gipson quote on p. 239).

19. Bessey quoted in Niels H. Hansen, "The Western Sand Cherry," *South Dakota Agricultural Experiment Station Bulletin* 87 (June 1904): 10–11.

20. Hansen to Bessey, Brookings, August 4, 1900, reel 10, and February 7, 1906, reel 16, Bessey Papers.

21. Niels E. Hansen, "Questions and Answers on Fruit Culture," South Dakota Agricultural Experiment Station Circular 35 (June 1941): 28.

22. Hansen, "Western Sand Cherry," 16, 33.

23. Niels E. Hansen, "Plums in South Dakota," *South Dakota Agricultural Experiment Station Bulletin* 93 (May 1905): 48, 65.

24. Ibid., 70–71; Budd and Hansen, *Manual*, 1: 66–92.

25. Hansen, "Plums in South Dakota," 68–69.

26. Hansen to Bessey, Brookings, March 27, 1907, and Bessey to Hansen, Lincoln, April 3, 1907, reel 18; Bessey to Hansen, Lincoln, July 16, 1908, reel 20, all in Bessey Papers.

27. Hansen, "Plant Introductions," 24.

28. Ibid., 26; *Field and Farm* 26, no. 1349 (December 9, 1911): 2.

29. Jules Sandoz, "A Message from the Wizard of Horticulture for the Nebraska Sand Hills," Nebraska State Horticultural Society Annual Report 52 (1921): 195.

30. Quoted in Bailey, *Native Fruits*, 426.

31. Niels E. Hansen, "Breeding Hardy Strawberries," *South Dakota Agricultural Experiment Station Bulletin* 103 (June 1907): 225, 232; William P. Kirkwood, "The Romantic Story of a Scientist: The Work of Professor Hansen in Discovering and Inventing Fruits and Forage That Will Stand Sub-Zero Weather," *The World's Work* 15, no. 6 (April 1908): 10115.

32. Hansen, "Plant Introductions," 43, 46 (quote on p. 43); "Hansen's New Hybrid Raspberry," *Field and Farm* 23, no. 1169 (May 23, 1908): 3; "South Dakota Reports," South Dakota Horticultural Society Annual Report 6 (1909): 205.

33. John S. Robertson, "Fruit Culture 1923 in the Southern Black Hills," South Dakota Horticultural Society Annual Report 21 (1924): 116; Harry R. Woodward, "A Great Fruit Grower and His Contributions" (typescript, Brookings: South Dakota State University, 1941), section b, p. 6.

34. Niels E. Hansen, "Secretary's Report," South Dakota Horticultural Society Annual Report 7 (1904): 17, and 11 (1908): 7; Hansen, "Civic Improvement," South Dakota Horticultural Society Annual Report 8 (1905): 108; Hansen, "The Silver Lining to the Horticultural Cloud," South Dakota Horticultural Society Annual Report 8 (1905): 172.

35. "Horticultural Program for South Dakota," February 1936, and "An Action Program for Horticulture in Wartime," 1941, both in carton 2, Hansen Papers.

36. Niels E. Hansen, "Fifty Years Work as Agricultural Explorer and Plant Breeder," *Iowa State Horticultural Society Transactions* 79 (1944): 31; Miller, "Eminent Horticulturalist," 278–279.

37. Niels E. Hansen, "A Hedge Plant from Siberia," South Dakota State Horticultural Society Annual Report 14 (1917): 212.

38. Accession log, 1930s (manuscript), Cheyenne Horticultural Field Station, Cheyenne.

39. "Hansen Commemoration," Brookings, August 25, 1949, carton 2, Hansen Papers.

10

Federal Engagement in Horticulture

With scarcity of water the most fundamental fact about the High Plains, it is not surprising that, as the region developed, larger, more complex, and far more expensive irrigation works were required. John Wesley Powell had foreseen that need in 1878. While he had recommended new laws and regulations for the arid West, he could not possibly have imagined the forthcoming massive federal financial and technical assistance, beginning with the Reclamation Act of 1902 and continuing to this day.

Passage of such an act had been deferred for twenty-four years because of state, territorial, and local resistance to any federal role in the distribution of irrigable water. In 1891, for example, Wyoming state engineer Elwood Mead had obtained a resolution from the National Irrigation Congress meeting in Salt Lake City, which asked the U.S. Congress to cede all public lands in the West to the states and to earmark those lands for reclamation by irrigation. Wyoming senator Francis E. Warren introduced, but failed to obtain, passage of legislation to do just that. In 1894 his Wyoming colleague, Joseph M. Carey, successfully obtained approval of an act that transferred to Wyoming and ten other western state up to 1 million acres of "desert lands," each for reclamation, on condition that those lands be sold to settlers in lots not exceeding 160 acres, of which 20 acres had to be irrigated within ten years.[1]

The Carey Act only marginally affected the High Plains, except around Wheatland in eastern Wyoming. Congress, meanwhile, had taken other steps toward reclamation by authorizing both an "Irrigation Survey" (1888) of the arid West under Major Powell and a more specific study (1896) of potential reservoir sites in Colorado and Wyoming. The study, conducted under the direction of Captain Hiram M. Chittenden of the Army Corps of Engineers, recommended five sites and, for the first time in writing, recommended that the reservoirs be constructed at federal expense. Again, Senator Warren introduced legislation, this time to appropriate monies for the construction of three reservoirs in Wyoming and one in Colorado; it, too, failed. In its place, Congressman Francis G. Newlands, who had failed to obtain reservoir legislation for his home state of Nevada, introduced an omnibus bill known as the Reclamation Act of 1902. Supported and signed into law by President Theodore Roosevelt, the act authorized the secretary of the interior to construct storage reservoirs as well as canals and other means for water distribution in each of sixteen western states and territories and to pay for them through the sale of public lands in the respective states. In time, federal appropriations actually covered all construction expenses, while the states retained control over the distribution of water. Bernard DeVoto's famous quip, "get out, and give us more money," still describes the prevailing western attitude toward the federal government.[2]

The first construction project under the Reclamation Act to affect the High Plains was Pathfinder Dam and Reservoir on the upper North Platte River in south-central Wyoming. Completed in 1909, Pathfinder had the capacity to store over 1 million acre-feet of water (an acre-foot is a unit of volume one acre in area and one foot in depth) for farmers, ranchers, and communities in eastern Wyoming and western Nebraska. By far the most spectacular reclamation project, with momentous impact on the development of the Front Range, was the diversion of Colorado River waters from the Western Slope to the South Platte River basin. Construction began during the New Deal in 1938 and was completed in 1959 with reservoirs, canals, and a thirteen-mile tunnel under the Continental Divide in Rocky Mountain National Park.

The Reclamation Act of 1902 enabled the federal government to carry out Powell's plan for providing irrigable water to the arid West, but the act made no allowance for federal support of the installations and services—the infrastructure—to lessen the risks to the yeoman farmers who were supposed to benefit from irrigation. The Smith-Lever Act of 1914 provided the means for the federal government to carry out the community-building aspect of Powell's plan.[3]

Essentially, Smith-Lever regularized and augmented the extension education services sponsored since the 1880s by the land-grant colleges through their agricultural experiment stations and made those services part of a comprehensive approach to improving rural life. The two most distinctive elements of extension education, the county agent and the demonstration farm, came into being during the decade before passage of Smith-Lever.

Although experiment station employees such as James E. Payne had been working in fields and gardens side by side with homesteaders since the 1890s, Colorado Agricultural College employed its first true county agent in 1912, financed cooperatively by county and federal governments. The agent, D. C. Bascom, who taught agriculture at Logan County High School (Sterling), was appointed only six years after the creation of the nation's first county agent position in Smith County, Texas.[4]

At the time, the creation of county agents' chief proponent, Seaman A. Knapp, was consulting with the United States Department of Agriculture (USDA) in Texas about eliminating the boll weevil from cotton fields. He is also credited with establishing the first private demonstration farm that was not part of a public experiment station in Terrell, Texas, in 1903. Knapp recognized farmers' inherent traditionalism and their fear of trying new techniques, so the novelty of his approach was to secure organizational partners (later, the federal government) that would share risk with the farmers. Ever focused on the applications of science to agriculture and the need for the experts to work in close association with those using their services, Knapp is best remembered for this statement: "What a man hears, he may doubt; what he sees, he may possibly doubt; but what he does, he cannot doubt." This aphorism remains true to this day in describing the ideal "cooperative" relationship between the land-grant colleges and rural residents of the High Plains.

The intellectual foundation of this relationship was expressed by the Country Life Commission (1908–1909) as part of the Progressive agenda, meant to "curb the arrogance of organized wealth and the wretchedness of poverty amid plenty."[5] Once derided as the work of effete, idealistic easterners seeking to restore the charms of rural life, the Country Life Commission has recently been recognized for what it was—namely, the first volley in a national effort to develop "a system of self-sustaining agriculture." The commission contrasted farmers committed to "scientific and self-perpetuating agriculture" with those farmers who move to new land when their current land begins to lessen in productivity or with those who remain on the land, do nothing to preserve the fertility of the soil, and end their careers in poverty and degradation. Anyone contemplating restoring, or already working to restore, the vitality of rural communities on the High Plains would do well to review the commission's slim, eminently readable report to Congress and the president.[6]

To chair the seven-member commission, funded by a $5,000 grant from the Russell Sage Foundation, President Roosevelt appointed Liberty Hyde Bailey (1858–1954), preeminent American horticulturist and former dean of agriculture at Cornell University, the land-grant college of the state of New York. In his letter of appointment, dated August 10, 1908, the president wrote that "agriculture is not the whole of country life. The great rural interests are human interests, and good crops are of little value to the farmer unless they open the door to a good kind of life on the farm."[7] His non-economic emphasis irritated many rural residents. Give us more money, and we can take care of our own uplift! *Field and Farm* suggested editorially that rather than engage in non-essential matters, the federal government simply needed to "take the big foot off the farmer's neck."[8]

The commission mailed out about 550,000 questionnaires to rural residents and received back 115,000. In addition, between November 9 and December 22, 1908, the commission held public hearings at thirty locations, including Cheyenne and Denver. Regrettably, no record remains of the luncheon conversation among members of the commission, the Wyoming State Board of Horticulture, Aven Nelson, and Governor B. B. Brooks at the governor's residence in Cheyenne on December 5, 1908. We do get a hint, however, that Nelson would have been very sympathetic to the commission's mission. He considered himself a disciple of Liberty Hyde Bailey, and the two men maintained professional ties for over half a century. Nelson's other mentor, Charles Bessey, later served on the Nebraska Rural Life Commission, state successor to the Commission on Country Life.[9]

So here we have three leading academics, each with his feet on the ground—Bailey, Bessey, Nelson—who, beyond their specialized interests, embraced a holistic approach to building a "[rural] civilization in full harmony with the best American ideals."[10] Of the commission's three fundamental recommendations—an exhaustive survey of rural life, the organization of nationalized extension, and a publicity campaign for rural progress—the professors would have been most keen on strengthening college extension education.

The Smith-Lever Act, signed by President Woodrow Wilson on May 8, 1914, literally institutionalized the approach of "taking the university to the people." The act formally authorized cooperation among the land-grant colleges, county governments, and the USDA "to aid in diffusing among the people of the United States useful and practical information on subjects relating to agriculture and home economics, and to encourage the application of the same." In mandating "instruction and practical demonstrations . . . to persons not attending or resident in said colleges," the act generally defined the role of the extension agent.[11]

By specifying home economics, the act responded to the fundamental question posed by President Roosevelt to the Country Life Commission: "[H]ow can the life of the farm be made less solitary, fuller of opportunity, freer from drudgery, more comfortable, happier, and more attractive?" As part of its charge, the commission "made a special effort to ascertain the condition of women on the farm" and concluded that "the relief to farm women must come through a general elevation of country living."[12]

The outbreak of World War I in 1914, and the U.S. entrance three years later, provided unanticipated urgency to the development of the newly organized Cooperative Extension Service (CES). The first order of business was agricultural, to help increase production of basic crops, especially wheat; the second order was horticultural, to support the "wise use and preservation of food and the raising of additional food for home use in gardens."[13]

By 1917, Colorado CES had placed permanent agents in most counties on the eastern plains and supported these generalists with specialists traveling from the Fort Collins campus. With a majority of Coloradoans living in cities, Colorado CES exercised the foresight of designing its programs for both urban and rural residents. No better example of that approach exists than its work in horticulture, promoting the establishment of vegetable gardens.

Toward that end, Colorado CES divided vegetable gardens into three types: the home garden, the vacant-lot garden, and the large-tract garden. While the home garden was the most common and incidentally ensured "the greatest cooperation between parents and children," the vacant-lot garden made use of land otherwise wasted and contributed to beautifying the community. Before his death in 1918, Mayor Speer directed city crews to plow vacant lots and arranged for Denver Water to provide free irrigation for Victory Gardens. Part of the wartime Food for Freedom campaign, Victory Gardens, like the earlier Pingree Gardens, served both aesthetic and humanitarian purposes, in particular providing fresh produce to some of Denver's poorest families. In Fort Collins, meanwhile, citizens waged a wartime campaign to turn every backyard into a vegetable garden and planted over 100 vacant lots to vegetables and grains. The same occurred in Greeley, where, in addition,

Victory Garden during World War I, Denver. Courtesy, Denver Public Library.

since 1915, members of the newly formed garden club volunteered as teachers' aides, showing children how to grow vegetables and helping them sell their produce at school markets. Indeed, during the time of war, rural youth on the High Plains necessarily devoted much of their extracurricular time to growing food.[14]

As to the large-tract type of garden, the municipality of Sterling, for example, made available a 7.5-acre tract to members of the local garden club; the school district underwrote the summer stipend for the agriculture teacher/ county agent to supervise the gardeners. As part of its expanded mission, Colorado CES not only assisted communities in the mechanics of starting new garden clubs but also provided gardening and canning specialists to help with practical instruction locally. In addition, under Smith-Lever, CES provided staff in support of youth activities under the now familiar symbol of the four-leaf clover, with the "H" on each leaf standing for "Head, Heart, Hands, and Health."[15]

As the CES developed into the principal vehicle to transmit useful knowledge from land-grant colleges to rural residents, the agricultural experiment stations remained the primary source of such knowledge. Because of the widespread planting of the potato in gardens as well as fields, and in response to crop failures recurring between 1911 and 1915, the Colorado Legislature appropriated special monies toward establishment of the Potato Experiment Station near Greeley. A cooperative undertaking of the Weld County commissioners, the Colorado Agricultural Experiment Station, and the USDA Bureau of Plant Industry, the station embarked upon research on diseases and cultivation techniques.

While early investigators failed to isolate the exact cause of the potato failures, in later years they would determine as culprit psyllid yellows, a disease induced by a toxin injected into potato leaves by the nymph of the potato psyllid (*Paratrioza cockerelli*). Beginning in 1918, their experiments came under the supervision of William O. Edmondson (1891–1961), horticulture and forestry graduate of Colorado Agricultural College, with prior experience as a high school teacher and county extension agent. His most notable experiments showed that potatoes receiving early irrigation sustained vigorous vine growth throughout the season and produced much larger yields of quality tubers than occurred under the more traditional practice of withholding water until potato vines showed wilting from lack of moisture.[16]

Determining the wilting point of plants was considered fundamental to understanding their minimal water requirements and became the subject of pioneering research at the USDA Field Station near Akron. "Wilting point" meant that point at which a plant could extract no more water from the soil. Using the sunflower as test plant, Lyman J. Briggs and H. L. Schantz found that the wilting point varied from 2 percent water content in sandy soils to 14 percent water content in silty clay loams and clay loams. Their laboratory research, combined with field trials, led to the development of new techniques for improving water intake by plants, by eliminating weeds and reducing evaporation. Their work resulted in better-informed selection of dry-land crops, most immediately favoring certain varieties of millet and wheat. As part of the wartime effort, wheat acreage in Washington County (Akron) increased from 31,000 acres in 1917 to 175,403 acres in 1920.[17]

Unquestionably, the push to increase food supplies added a sense of urgency to the production, collection, and distribution of high-quality seeds. Because of the difficulty of obtaining seeds from overseas during wartime, additional eastern seed houses joined companies such as D. M. Ferry in contracting with farmers along the Arkansas River in southeastern Colorado to grow both field and garden seeds under their supervision.[18]

Along the Front Range and in northeastern Colorado, Frederick C. Vetting traveled as a salesman for eastern seed houses, selling garden seeds at

Rocky Mountain Seed Company, Denver, 1920s. Courtesy, owners.

five cents per package. Vetting had emigrated from Wisconsin to Colorado in 1906, attracted to a better climate for his pulmonary condition. By 1920, local farmers and home gardeners had convinced Vetting to start his own seed house in Denver. The Rocky Mountain Seed Company would earn the reputation as the most reliable retail and wholesale seed house for farmers, ranchers, and gardeners—both urban and rural—on the High Plains.[19]

On April 10, 1917, Colorado became the first state on the High Plains to adopt a pure seed law, and the state established a seed-testing laboratory at Colorado Agricultural College. The new law required labeling of the contents (species and variety) of every seed packet sold in Colorado, percentage of purity, rate of germination, date of testing, where the seed was grown, and the presence (species and quantity) of noxious weeds. The law did not prohibit the sale or purchase of inferior seed.

While the new seed law provided for farmers and gardeners to send their seeds to the college for testing at no cost to them, the legislature had declined to provide funding for inspection and enforcement. This meant that unscrupulous dealers continued to distribute large quantities of impure seed and

Rocky Mountain Seed Company, Denver, 1920s. Courtesy, owners.

caused the state to spend large sums unnecessarily on eradicating noxious weeds. In time, the 1917 seed law would be repealed, reenacted, amended, and adopted, most recently in 1993 as the Colorado Seed Act, closely following the Recommended Uniform State Seed Law.[20]

By 1914, with the urbanization of the Front Range, most major commercial orchards had been relocated to the Western Slope, leaving only small orchards of home and hobby value. Commercial orchardists had convinced the legislature in 1914 to fund the position of state horticulturist. Its first occupant was Emil Peter Sandsten (1868–1961), who had arrived a year earlier from the University of Wisconsin to take the position of professor of horticulture at Fort Collins.[21] Although he did his major work on the Western Slope, Sandsten oversaw the development of a small experimental orchard, located on benchland four miles east of Fort Collins. There, college faculty and students tested the relative hardiness and adaptability of different varieties of fruit trees meant primarily for homesteads on the eastern plains. Their findings were consistent with contemporary trials at the Hays station in western

Kansas and the Sandoz station in northwestern Nebraska: establish windbreaks around orchard plots in advance of planting fruit trees; plant two- and three-year-old fruit trees, not year-old trees; where possible, plant fruit trees at least twenty-four feet apart; immediately after planting, prune fruit trees so branches correspond with reduced root areas; plant cover crops around trees; heavily irrigate once during the growing season, then again once in late fall.[22]

Although it is unclear whether staff at the Fort Collins orchard still cultivated with horse-drawn plows, it was during the World War I period that tractors supplanted horses as the supplier of moving power. Tractors became common in pulling agricultural implements, first in dry-land and then in irrigated farming. As smaller tractors became permanently fitted with certain implements, they could be used as well in gardening and ornamental horticulture. County agents provided much of the information and arguments for purchase of these labor-saving machines.

During the 1920s, nurserymen and other commercial horticulturists began using relatively small rotary tillers, or "earth grinders." The machines were invented in Switzerland around 1910, and American manufacture of the machines did not start until the late 1930s. For home garden use, production of the popular, relatively inexpensive, gasoline-powered "Rototiller" did not occur until after World War II.[23]

Meanwhile, by making it easier to bring larger and larger plots of dry land under cultivation, new machinery contributed to, but was by no means the primary cause of, soil depletion on the High Plains. At the same time, those devices were harnessed for the purpose of soil conservation. Outstanding among conservation activities on the High Plains during the 1920s was federal support of seedling nurseries and tree distribution to landowners through the Cooperative Farm Forestry Act of 1924, sponsored by Representative John D. Clarke of New York and Senator Charles L. McNary of Oregon.

While prior forestry legislation had provided for some technical and financial assistance, Clarke-McNary accelerated and broadened that assistance, authorizing the U.S. Forest Service to cooperate with the states in assisting private landowners to plant windbreaks and shelterbelts. Since 1924 the Forest Service has made millions of seedlings available to landowners on the High Plains at little or no cost. For the first decades, stock came from three principal sources: the USDA Northern Great Plains Field Station in Mandan, North Dakota, the Bessey Nursery in the Nebraska National Forest, and the Kansas Agricultural Experiment Station in Hays.

As noted, nursery stock had been grown at Hays since 1912, but scientific research on trees and shrubs occurred only after the appointment of E. W.

Johnson as state forester in 1927. A graduate of Colorado Agricultural College in horticulture and forestry, Johnson managed trial blocks for testing forest trees and shrubs, as well as fruit trees, for their adaptability to western Kansas; he also served as extension forester and handled the station's horticultural correspondence.[24]

Johnson's primary charge was to oversee the propagation, cultivation, and at-cost sale of seedlings, transplants, and shrubs. During his first year, station staff distributed approximately 10,000 seedlings under Clarke-McNary; later, annual distribution reached as high as 868,000. Recommended species were red cedar (*Juniperus virginiana* L.), bur oak (*Quercus macrocarpa* Michx.), American elm (*Ulmus americana* L.), Chinese elm (*Ulmus parvifolia* Jacq.), and cottonwood (likely *Populus angustifolia* James and *Populus deltoides* Bartr. ex Marsh.). In 1936 the station made distributions for field planting (59.3%), farmstead windbreaks (34.3%), farm woodlots (5.6%), and rural school grounds (0.8%).[25]

Over a quarter century, the Hays station distributed more than 9 million trees, mostly to western Kansas, of which half a million went to the eight northwestern counties. Then, in 1951, State Attorney General Harold R. Fatzer ruled that any distribution by the station for commercial sale that was not the by-product of experimentation was a violation of the Kansas Constitution. Not surprisingly, this ruling resulted from years of pressure by commercial nurserymen and officers of their association who claimed the state was competing against private business.[26]

In Colorado, meanwhile, the USDA Bureau of Plant Industry and the Colorado Agricultural Experiment Station had conducted cooperative forest experiments at Akron since 1908. Colorado state forester W. J. Morrill used this research in preparing "Trees for Non-Irrigated Regions in Eastern Colorado," with his suggestions for planting methods and required maintenance. Six years before the passage of Clarke-McNary, the Forestry Department at Colorado Agricultural College started distributing up to 100,000 seedlings annually at cost of production to residents of eastern Colorado. Of the twelve species available, Morrill favored black locust (*Robinia pseudoacacia* L.) and honey locust (*Gleditsia triacanthos* L.) for posts, fuel, and farm repairs; Russian-olive (*Elaeagnus angustifolia* L.), jack pine (*Pinus banksiana* Lamb.), and red cedar (*Juniperus virginiana* L.) for windbreaks; and hackberry (*Celtis occidentalis* L.), ponderosa pine (*Pinus ponderosa* Laws.), and, where water was readily available, cottonwood (*Populus angustifolia* James) for yard planting. While some seedlings may have been grown at Akron, most came from federal nurseries in the region.[27]

Windbreak, planted 1920s near LaGrange, Wyoming. Courtesy, Ronald K. Hansen.

Besides underwriting part of the Colorado tree-planting program, Clarke-McNary contributed to interagency and regional coordination. For example, during the summer of 1931, representatives of four separate agencies participated in an inspection tour through eastern Colorado: USFS regional forester Fred R. Johnson (Denver), Kansas state forester E. W. Johnson (Hays), Colorado extension forester R. E. Ford (Fort Collins), and windbreak specialist John L. Emerson, Cheyenne Horticultural Field Station. In general terms, the inspection team checked on the use made of trees furnished by the state forester and whether such use benefited landowners and the state and federal governments; the team also assessed the health of different varieties of trees and shrubs and addressed procedural questions regarding Clarke-McNary.

Starting at the Akron station, the team visited three farms about thirty miles north where it found that the ponderosa pine remained the outstanding choice for growth, hardiness, and drought tolerance. In Lincoln County (east of Limon), the team found that the county agent had diligently sought to advise farmers on proper shelterbelt planting. Too often, he noted, trees were planted in the wrong places: "In many instances trees are set too close to buildings and there is no snow trap, as result of which the snow banks up against the buildings. In other cases, groves have been established east of

Windbreak, planted 1920s near LaGrange, Wyoming. Courtesy, Ronald K. Hansen.

buildings where they do little good when protection is most needed." In other counties on the eastern plains, the team noted very little planting despite the urgent need to reduce soil blowing. While county agents expressed interest in the matter, the team believed someone needed to encourage and train them, leading to a recommendation to hire a regional extension forester. As with any bureaucracy, CES housed some employees who lacked both commitment and vitality to execute its noble mission.[28]

Other employees, however, were indefatigable, among them W. O. Edmondson at the Greeley Potato Experiment Station. In 1929 he moved to Laramie to become Wyoming's first full-time extension specialist in forestry and horticulture. Concerning his first year in Wyoming, Edmondson recorded that, as a result of participating in seventy-eight community meetings, he successfully distributed 50,000 seedlings under Clarke-McNary. During his thirty-two-year tenure at Wyoming, he distributed over 1 million seedlings to the four eastern plains counties alone. The most popular varieties: ponderosa pine, Chinese elm (*Ulmus parvifolia* Jacq.), Russian-olive, and Siberian pea (*Caragana arborescens* Lam.). All seedlings came from the USDA Northern Plains Station at Mandan.[29]

While Edmondson served as emissary of horticulture in all its forms, he viewed his main task as helping farmers and ranchers plant shelterbelts.

Federal Engagement in Horticulture   183

Windbreak, planted 1920s near Lingle, Wyoming. Courtesy, Ronald K. Hansen.

Progress occurred in small steps. In 1931 he joined the local county agent in helping five landowners in Platte County (Wheatland) plant shelterbelts, provided fourteen others with trees, and gave one demonstration-lecture on transplanting evergreens. Apparently, general landowner interest did not materialize. To further homestead improvement, Edmondson persuaded W. C. Deming, Cheyenne pioneer and former University of Wyoming trustee, to establish a fund "for prizes to owners of successful farm shelterbelts and well-improved farm yards." The idea was for CES to solicit and review applications annually and make awards to farmers with the best dry-land and the best irrigated shelterbelts. For reasons unclear, the awards ceased after six years.[30]

By the time Clarke-McNary became law in 1924, the Bessey Nursery had already distributed 1.9 million seedlings to Nebraska residents west of the 100th meridian, as provided by the Kinkaid Act of 1904. Clarke-McNary then absorbed and expanded the earlier provision. Nebraska CES served as the primary agency to ensure the procurement and distribution of trees under Clarke-McNary. From 1926 through 1950, western Nebraska received more than 24.5 million seedlings for windbreaks, shelterbelts, and woodlots. Nearly half of those seedlings came from the Bessey Nursery.[31]

In the Bessey tradition, University of Nebraska faculty and staff did much to establish Nebraska's reputation as the tree-planting state. As in Colorado and Kansas, the Nebraska state forester was a university employee in the high-profile Conservation and Survey Division, which, among other activities, promoted tree planting. In 1935 the division published a list of trees recommended for the various districts of Nebraska. For the western district, it recommended honey locust, American elm, green ash (*Fraxinus pennsylvanica* Marsh.), black walnut (*Juglans nigra* L.), boxelder (*Acer negundo* L.), cottonwood, Norway [Carolina] poplar (*Populus x brayshawii* Boivin), white spruce (*Picea glauca* [Moench] Voss), Koster's blue spruce (*Picea pungens* Engelm. cv. "Kosterana"), blue spruce (*Picea pungens* Engelm.), jack pine (*Pinus banksiana* Lamb.), ponderosa pine, and Austrian pine (*Pinus nigra* Arnold). That same year (1935) the university's extension forester took leave to direct the Nebraska portion of the Prairie States Shelter Belt Project, an ambitious New Deal program to assist private landowners on a 100-mile-wide swath of the Great Plains from Canada to Texas, but mostly east of the High Plains.[32]

During the late 1920s, Nebraska governor Adam McMullen brought together representatives of government agencies, businesses, and civic organizations in a statewide committee to plan and organize a comprehensive tree-planting campaign. His successor, Governor Arthur J. Weaver, expanded the committee's purpose to include planting of flowers, vines, and shrubs; renamed it the State Committee on Tree Planting and Landscape Beautification; and proclaimed the entire week before Arbor Day 1929 "Tree Planting and Landscape Beautification Week."[33]

The mastermind behind this statewide beautification effort, and committee chair, was George E. Condra (1869–1958), director of the Conservation and Survey Division. A onetime college athlete who earned a Ph.D. in geology from Cornell University, Condra had joined the Nebraska faculty in 1902, learned to master academic politics, served as confidential adviser to governors, and gained the reputation as the state's leading advocate for conservation in the tradition of Theodore Roosevelt. His plan was to engage citizens in every walk of life in conservation efforts at little or no expense to government, but evidence of the plan's successful implementation appears lacking. Somehow, hierarchically organized campaigns dampen voluntary enthusiasm and effort, which is usually local and egalitarian.[34]

Generally speaking, government is least well suited for contributing to quality-of-life amenities and most effective when addressing the basic needs of food, shelter, and security for its citizens. The latter is precisely what the federal government, under the leadership of President Franklin D. Roosevelt, sought to provide through the New Deal in wake of the calamities of the late 1920s and early 1930s. Although only indirectly affecting horticulture, the

Agricultural Adjustment Act (AAA) of May 12, 1933, was arguably the most important effort to restore and stabilize agriculture and communities on the High Plains. Without abandoning its instructional mission, the Cooperative Extension Service took up the task of administering the AAA. Essentially, county extension agents oversaw the reduction in planting of basic crops; farmers, in return, received guaranteed prices. Although the act was declared unconstitutional in 1936, then amended and modified on numerous occasions, the principle of adjusting agricultural production to demand continued in effect into the 1990s.[35]

Agricultural adjustment, however, meant more than production control. In the wake of the dust storms of the 1930s, the AAA sought to tie cultivation to intelligent soil management, which led to enactment of the Soil Conservation and Domestic Allotment Act in 1936. This act provided for a permanent program to control and prevent soil erosion and directed the secretary of agriculture to establish the Soil Conservation Service, now known as the Natural Resource Conservation Service, to carry out national restoration and conservation policies. Secretary of Agriculture Henry A. Wallace recommended the establishment of local soil conservation districts as the most effective way to administer the act and to voluntarily engage private landowners and other users of private land. The secretary provided all governors with a model act, with Colorado the first High Plains state to adopt soil conservation districts in 1937. The Soil Conservation Service provided technical assistance to local districts, while the CES provided educational services; the two agencies did not always work together happily.[36]

The most popular New Deal program to benefit homesteads and smaller communities on the High Plains was rural electrification. Although county extension agents had been assisting farmers and ranchers in electrifying their operations since the mid-1920s, the Rural Electrification Administration Act of 1936 provided for loans and other forms of assistance to associations and nonprofit corporations that furnished electricity to rural areas. In 1935, less than 10 percent of the rural High Plains was electrified; by 1945, that number was roughly 50 percent. The effect of rural electrification on horticulture and rural life in general would be incalculable.

Year after year, from 1914 through 1945, the agricultural experiment stations through the Cooperative Extension Service helped enlighten both rural and urban residents on the value of home gardening. Colorado employed its first extension specialist in horticulture in 1929. In his inaugural publication, A. M. Binkley reiterated the argument that the vegetable garden was useful, indeed suitable, to all families by reducing the food bill; providing fresh, high-quality vegetables in season; supplying vegetables for canning, drying, and winter storage; enabling substitution of vegetables for more expensive foods; providing the most healthful diet possible; and, for the urban family, providing as well "a place for pleasant outdoor recreation."[37]

In Wyoming, meanwhile, W. O. Edmondson spent much time in the southeastern counties, where he inaugurated a potato certification service for commercial growers and worked in a more general capacity with market and home gardeners. In 1930 he served as horticultural specialist to the Goshen County (Torrington) Farm and Home Economics Conference, helping to plan major increases in farm production; subsequently, he served as secretary to the horticulture committee, which prepared recommendations on fruit varieties most likely to thrive in Goshen County.[38]

Edmondson extolled both the economic and non-economic value of horticulture. A few fruit trees and small fruits, he wrote in 1935, should be part of everyone's garden plan, whether on the farm or ranch or in the town or city. He urged Wyoming residents to eat more fruit for health's sake, as well as because fruit adds variety and flavor to other foods. Additionally, fruit trees create attractive shade, and small fruits do well as hedges or shelterbelts.[39]

From a pioneer ranch woman who lived along Horse Creek on the eastern plains, Edmondson received this comment in 1930: "I appreciate my yard of shrubs and flowers more than ever these days. When things go wrong around the dining table or the children become unruly, I get quite a bit of satisfaction by going out of the house into my flower garden and looking at the growing things. I see the flower heads lifted up toward heaven always striving for light and doing their best to develop. Such a sight . . . gives me renewed vigor and strength to return to the house and care for the family."[40] That sentiment was heard again and again throughout the High Plains, in both fiction and nonfiction.

While the Nebraska Agricultural Experiment Station had recognized the Jules Sandoz orchards as a substation, experiments with vegetables in northwestern Nebraska took place at the Scottsbluff substation, but apparently not until the late 1920s. The result of yield tests (1931–1936) made on those vegetable varieties considered most likely to thrive with limited or no irrigation would not surprise today's home gardener at the lower elevations of the High Plains: peas and beans in the pea (Leguminosae) family; carrots, parsnip, celery, and parsley (Umbelliferae); potatoes, eggplant, peppers, and tomatoes (Solanaceae or nightshade); cucumbers, melons, squash, and pumpkins (Cucurbitaceae or gourd); cabbage, cauliflower, turnip, radish, kohlrabi, rutabaga, and Chinese cabbage (Cruciferae or mustard); sweet corn

(Gramineae); onion, garlic, and leek (Amaryllidaceae or lily); red beets, spinach, and chard (Chenopodium or goosefoot); and lettuce, salsify, and endive (Compositae).[41]

Of all the garden vegetables, the tomato was, and remains, the favorite and the most difficult to grow. To begin with, the tomato is a non-native, warm-season plant, but when the temperature reaches about 85°F, the blooms tend to fall off so the fruit fails to set. Since the tomato is frost intolerant, its seed must be started in a protected atmosphere; allowed to "harden off," either by gradually reducing the temperature of the hotbed or greenhouse or by increasing exposure to wind and sunlight; and, finally, transplanted into garden or field after all danger of frost is past. While recent innovations make it possible to protect some varieties of tomatoes from light frost, that was not the case in western Nebraska in the 1930s.[42]

During the late 1930s, the Nebraska Agricultural Experiment Station, through its extension horticulturists, engaged both home gardeners and commercial growers in statewide testing of five promising new tomato varieties, all developed in the laboratories of land-grant colleges. The aim of the test was not so much to help scientists by providing performance results as it was to demonstrate to Nebraska families that, despite the capricious climate, tomatoes could still be grown for home use. Testing was administered so as to engage citizens at every locale and, in the process, to make the results of research useful everywhere.

In response to publicity through local newspapers, interested parties applied to the CES, and county agents selected ten to fifteen participants ("cooperators") per county. Once participants were chosen, the agents distributed the five new seed varieties as well as seeds of two standard varieties for comparison purposes, accompanied by a circular that described the new varieties, their potential adaptability to local conditions, and suggestions for how best to cultivate the tomatoes. They also provided forms for cooperators to record soil conditions, past history of test plots, descriptions of tomato cultivation, tomato yields, and their conclusions concerning the best variety. Throughout the growing season, cooperators received additional circulars and had the opportunity to attend regional meetings to discuss their progress. In addition, agents requested that cooperators save enough seed to supply themselves and their neighbors for the next season. Altogether, over 400 out of an initial 1,000 cooperators completed their test reports. Cooperators generally favored the Bison variety, developed by Albert F. Yeager at the North Dakota Agricultural Experiment Station, primarily because its fruit set best during the hot and dry conditions of July and August, so prevalent on the High Plains.[43]

With the coming of World War II, as during World War I, the CES served as principal administrator of federal programs meant to increase food production. County and home demonstration agents advised families about growing, conserving, and preserving their own food. Through local newspapers, special publications, seminars, and radio broadcasts, the CES publicized the latest techniques in canning, freezing, dehydrating, brining, and storing. Through farm and home visits, agents advised individuals in carrying out specific projects such as growing new varieties of vegetables and fruits and remodeling kitchens.[44]

Once again, Victory Gardens attracted the most public attention. "Plant a Victory Garden: Help Win the War!" The aim was for American families—urban, suburban, and rural—to grow a substantial portion of their total food supply so commercially grown produce could go to the Allied troops abroad. And, once again, the CES was given the responsibility to encourage Victory Gardens. In Nebraska, agents worked primarily through the Federated Garden Clubs of Nebraska, an offshoot of the Nebraska Horticultural Society but separately incorporated since 1935, and through county 4-H Clubs, successors to the boys' and girls' clubs of the 1910s and 1920s.[45]

In a wartime pep talk to garden club members in Nebraska, Mrs. Adam S. Wagner, the Rocky Mountain regional vice president of the National Garden Clubs, called for wasting no seeds, "not even one ten cent package of lettuce"; making certain that sufficient irrigation water would be available for Victory Gardens, especially when members used vacant town lots; properly handling the necessary insecticides and fertilizers; and anticipating the best methods for storing fresh produce.[46]

In addition to sponsoring Victory Gardens, many local garden clubs took over maintenance of roadside and town parks; planted trees, shrubs, and flowers at military installations, including prisoner-of-war camps; and provided hospitals and United Service Organizations with flowers. In addition, many commercial nurseries contributed the trees garden club members planted as living memorials to fallen soldiers.[47] Garden clubs reached the peak of their memberships, programs, and activities during World War II. While many clubs still exist, membership has declined, members tend to be older, and their activities tend to concentrate on ornamental horticulture.

The pressing need to increase the food supply for the United States and its wartime allies, coupled with the cumulative effect of many years of excessive plowing, overgrazing, soil blowing, and water erosion, placed a renewed sense of urgency on restoration and conservation of soil on the High Plains. At the very time soil conservation districts were established, the Hays station gained preeminence for its research on selecting and modifying native grasses. A significant, although secondary, result of that research was the creation of new varieties of drought-resistant and heat-tolerant grasses for lawns and other landscape areas.

Since the early 1930s, researchers at Hays had been testing varieties of grasses collected worldwide by USDA explorers. None of these foreign varieties, however, proved successful for widespread restoration use on the High Plains. As a result, Leon R. Wenger, forage crops specialist at Hays, turned to experimenting with two native grasses: buffalo grass (*Buchloë dactyloides* (Nutt.) Engelm.) and blue grama grass (*Bouteloua gracilis* (H.B.K.) Lag. ex Griffiths). Buffalo grass is the dominant "short-grass" of the High Plains, naturally growing up to eight inches tall, while blue grama grass dominates the driest prairie land, growing up to twelve inches tall; both grasses are perennials.

From 1937 until 1943, Wenger conducted breeding experiments to modify these native grasses so they could be grown economically for restoration and conservation purposes. Buffalo grass posed the greater challenge: its seeds (burs) develop naturally virtually at ground level, making collection difficult; its normal germination rate is very low; and it can lie dormant for four or five years, depending on drought and other climatic conditions.

In the late summer of 1937, Wenger collected his first seeds, planted one per small pot, placed the pots in a greenhouse over the winter, and set out the young plants in the spring. Not surprisingly, the plants showed different dominant characteristics: some spread rapidly by surface runners or stolons, which ranged from several inches to several feet in length; some took root at the nodes, where blades grew out of the stems, to produce new runners; some vegetated more than others, its usual way of propagating since buffalo grass is dioecious, bearing male and female flowers on different plants; and some plants grew up to ten inches tall while others barely reached one and one-half inches.

By the summer of 1940, Wenger's experiments involved more than 10,000 individual plants from South Dakota, Nebraska, Kansas, and Texas. He selected and bred for those characteristics most desirable for different uses—pastures, lawns, landscape areas, and, most immediately, new military airfields on the plains. Wenger released his most promising cultivar, "Hays buffalo grass," in 1942. A station committee of three, chaired by Wenger, developed standards for certifying cultivars as buffalo grass to farmers, nurseries, and government agencies.

Meanwhile, after several attempts at refitting existing machinery, mechanics at the station successfully rebuilt a small combine harvester. By attaching a remodeled cutting bar moving at ground level, they could efficiently harvest the buffalo grass. Although several commercial operators copied the invention, factory-built harvesters were not manufactured until the postwar era.

In the course of his experiments, Wenger concluded that buffalo grass seed required special treatment to break its dormancy and increase its germination rate: soak seeds in a one-half of 1 percent solution of saltpeter (potassium nitrate) for twenty-four hours, then chill at 40°F for six weeks, resoaking twice at two-week intervals, after which remove from cold storage and thoroughly dry. As a result, germination rates rose dramatically from roughly 7 to 80 percent. For the provision of large treating tanks, refrigeration units, and rapid drying equipment, Wenger relied on the Army Air Forces. In fact, the military eventually purchased more than 60,000 pounds of buffalo grass seed, enough to cover about 4,000 acres of airfields.[48]

In time, Wenger's federally supported research on buffalo grass would apply directly to ornamental horticulture. Replacing lawns of water-guzzling, non-native grasses with improved varieties of "water-wise" native buffalo grass is now considered the most efficient and least expensive way to reduce water consumption. Overall, however, during the New Deal and postwar eras, the principal source of horticultural improvements, and the outstanding example of federal engagement in horticulture on the High Plains, was the Cheyenne Horticultural Field Station.

Notes

1. Robert G. Dunbar, *Forging New Rights in Western Water* (Lincoln: University of Nebraska Press, 1983), 39; William L. Hewitt, "The 'Cowboyification' of Wyoming Agriculture," *Agricultural History* 76, no. 2 (Spring 2002): 492.

2. Dunbar, *Forging New Rights*, 50; Edward K. Muller, ed., *DeVoto's West: History, Conservation, and the Public Good* (Athens, Ohio: Swallow, 2005), 88.

3. Donald J. Pisani, "Reclamation and Social Engineering in the Progressive Era," *Agricultural History* 57 (January 1983): 63.

4. Robert G. Dunbar, "History of [Colorado] Agriculture," in Leroy Hafen, ed., *Colorado and Its People* (New York: Lewis Historical Publishing, 1948), 2: 156.

5. Samuel Eliot Morison, *The Oxford History of the American People* (New York: Oxford University Press, 1965), 811–812.

6. Liberty Hyde Bailey, ed., *Report of the Commission on Country Life* (New York: Sturgis and Walton, 1911), 84–85; Scott J. Peters and Paul A. Morgan, "The Country Life Commission: Reconsidering a Milestone in American Agricultural History," *Agricultural History* 78, no. 3 (Summer 2004): 292.

7. Bailey, *Report*, 42–43.

8. *Field and Farm* 23, no. 1197 (December 5, 1908): 4.

9. Minutes, December 5, 1908, Wyoming State Board of Horticulture Second Biennial Report, 8; Roger L. Williams, *Aven Nelson of Wyoming* (Boulder: Colorado Associated University Press, 1984), 292; Richard A. Overfield, *Science with Practice: Charles E. Bessey and the Maturing of American Botany* (Ames: Iowa State University Press, 1993), 168–169.

10. Bailey, *Report*, 24.

11. Act of May 8, 1914, *The Statutes at Large of the United States of America*, 63rd Congress, 2nd sess., Ch. 79, 372–375.

12. Bailey, *Report*, 42, 103, 105.

13. Wayne D. Rasmussen, *Taking the University to the People: Seventy-five Years of Cooperative Extension* (Ames: Iowa State University Press, 1989), 71–74.

14. M. E. Knapp, "County Agent's Report," Historical Development and Harvest Edition, *Weld County News* (1921): 32; Lyle W. Dorsett and Michael McCarthy, *The Queen City: A History of Denver*, 4th ed. (Boulder: Pruett, 1986), 144.

15. W. E. Vaplon, "Reports and Plans of Town and City Garden Clubs," *Colorado Agricultural College Extension Bulletin*, ser. 1, no. 133 (February 1918): 4–7, 14.

16. Alvin T. Steinel, *History of Agriculture in Colorado* (Fort Collins: Colorado State Agricultural College, 1926) 428–430.

17. Bentley W. Greb, "Significant Research Findings and Observations from the U.S. Central Great Plains Research Station and Colorado State University Experiment Station Cooperating, Akron, Colorado, Historical Summary, 1900–1981" (mimeograph, Akron, Colo.: U.S. Central Great Plains Research Center, 1981), 2; Lyman J. Briggs and H. L. Shantz, "The Water Requirements of Plants. I. Investigations in the Great Plains in 1910 and 1911," *USDA Bureau of Plant Industry Bulletin* 284 (Washington, D.C., 1913): 46–48; Dunbar, "History of Agriculture," 135.

18. *Field and Farm* 29, no. 1460 (January 24, 1914): 4, and 31, no. 1664 (December 29, 1917): 11.

19. Tom Noel, "Seed Store Firmly Planted in LoDo," *Rocky Mountain News*, March 1, 2003.

20. Steinel, *Agriculture in Colorado*, 578–579; E. P. Sandsten, "The Pure Seed Law in Colorado," Colorado Agricultural Experiment Station Annual Report (1937–1938): 5.

21. E.P. Sandsten, *First Biennial Report of State Horticulturalist* (Fort Collins, 1914), 4.

22. E. P. Sandsten and C. M. Tompkins, "Hardy Varieties of Apples for Northeastern Colorado," *Colorado Agricultural Experiment Station Bulletin* 292 (May 1924): 3.

23. "A Walking Cultivator," *Field and Farm* 31 (June 24, 1916): 2; "Gardening beyond the Plow" (leaflet, Troy, N.Y.: Garden Way, Inc., 1981).

24. "Fort Hays Branch Experimental Station Annual Report" (typescript, Hays: Western Kansas Agricultural Research Station, 1931) (hereafter cited as WKARC), 80–83.

25. E. R. Ware and Lloyd Smith, "Woodlands of Kansas," *Kansas State Agricultural Experiment Station Bulletin* 285 (July 1939): 9, 21, 24, 27.

26. William M. Phillips, "A History of the Agricultural Research Center–Hays: The First 100 Years," *Kansas State Agricultural Experiment Station Bulletin* 663 (April 2001): 69; excerpt from ruling by attorney general, September 25, 1951, appended to "Fort Hays Branch Experimental Station Annual Report," 1952.

27. W. J. Morrill, "Trees for Non-Irrigated Regions in Eastern Colorado," *Colorado Agricultural College Extension Bulletin*, ser. 1, no. 123 (September 1917): 3–20; Steinel, *Agriculture in Colorado*, 462.

28. Fred R. Johnson, memorandum, June 12, 1931, WKARC.

29. Edmondson obituary, *Laramie Boomerang*, May 19, 1961, 3; W. L. Quayle, "Trees—Wyoming's 25-Year Record," Wyoming Agricultural Experiment Station Circular 46 (September 1951): 6–7; George W. Boyd and Burton W. Marston, *The Wyoming Agricultural Extension Service and the People Who Made It, 1919–1964* (Laramie: University of Wyoming Publications, 1965), 44.

30. Quote from W. O. Edmondson, "Trees Improve Your Farm," Wyoming Agricultural Extension Service Circular 27 (February 1947): 28; Edmondson, "Trees for Protection and Profit," Wyoming Agricultural Extension Service Circular 116 (April 1951): 39; Frank P. Lane, "A History of Agricultural Extension Service in Wyoming Counties" (typescript, Laramie: University of Wyoming, 1964), 183.

31. Everett N. Dick, *Conquering the Great American Desert* (Lincoln: Nebraska State Historical Society, 1975), 135–137.

32. United States Forest Service, *Possibilities of Shelterbelt Planting in the Plains Region: A Study of Tree Planting for Protective and Ameliorative Purposes as Recently Begun in the Shelter Belt Zone of North and South Dakota, Nebraska, Kansas, Oklahoma, and Texas by the Forest Service* (Washington, D.C.: Government Printing Office, 1935), 54.

33. Raymond J. Pool, "Fifty Years on the Nebraska National Forest," *Nebraska History* 34 (September 1953): 174; George E. Condra, "Tree Planting and Landscape Beautification in Nebraska," *Conservation and Survey Division Bulletin* 2 (Lincoln, March 16, 1929): 5.

34. Vernon L. Souders, "Conservation and Survey Division Directors," http://snr.unl.edu/csd/history/directors.asp; accessed 2004.

35. Rasmussen, *Taking the University to the People*, 91, 96–97.

36. Ibid., 101–102.

37. A. M. Binkley, "The Home Vegetable Garden," *Colorado Agricultural Experiment Station Bulletin* 354 (February 1930): 3.

38. Albert E. Bowman, "History of the Agricultural Extension Work in Wyoming" (Laramie: Wyoming Agricultural Extension Service, 1964), 59–60; Alfonso F. Vass, "An Economic Survey of Goshen County, Wyoming," Wyoming Agricultural Extension Service Circular 29 (May 1930): 40.

39. W. O. Edmondson, "Fruit Raising in Wyoming," Wyoming Agricultural Extension Service Circular 58 (January 1935): 3.

40. Ibid., 5.

41. Lionel Harris, "Vegetable Variety Tests at the Scottsbluff Substation," *Nebraska Agricultural Experiment Station Bulletin* 300 (June 1936): 23.

42. H. O. Werner, "Varieties of Tomatoes Recommended for Various Localities in Nebraska on the Basis of Their Physiological Adaptations," Nebraska Horticultural Society Annual Report (1938): 27–29.

43. J. O. Dutt, "A Summary of New Tomato Variety Tests in 80 Nebraska Counties," Nebraska Horticultural Society Annual Report (1940): 47–54.

44. Rasmussen, *Taking the University to the People*, 109–112.

45. E. H. Hoppert, "The Victory Garden Program for Nebraska," Nebraska Horticultural Society Annual Report (1941): 37–41.

46. Mrs. Adam S. Wagner, [no title], Nebraska Horticultural Society Annual Report (1941): 4.

47. Ibid. (1941): 4, and (1944): 79.

48. Leland E. Call and Louis C. Aicher, "A History of the Fort Hays (Kansas) Branch Experiment Station 1901–1962," *Kansas State Agricultural Experiment Station Bulletin* 453 (May 1963): 83–87; Aicher, "Progress in the Improvement of Buffalo Grass for Western Kansas," 1938, 33, 36, WKARC; Aicher, "Buffalo Grass Improvement," 1940, 191–193, WKARC; "Report of Committee Appointed to Name the Improved Strain of Buffalo Grass; also to Suggest Standards for Certification," 1942–1945, 39–40, WKARC.

11

The Cheyenne Horticultural Field Station

On Sunday, October 11, 1936, President and Mrs. Franklin D. Roosevelt paid a short visit to the Cheyenne Horticultural Field Station. As it was an unusually warm and wind-free day, they arrived in a touring car with the top down and were greeted by Superintendent Aubrey C. Hildreth. The presidential party then proceeded to Roundtop, a landmark knoll on the station grounds, to greet Civilian Conservation Corps members at their campsite; then it was on to inspect Fort Warren, a mile away, and to participate in a reelection campaign rally at the parade grounds, attended by a crowd of 20,000—greater by 4,000 than the population of the City of Cheyenne. Mrs. Roosevelt graciously accepted a basket of large chrysanthemums; it was her fifty-second birthday. Although we found no record of what Dr. Hildreth discussed with the Roosevelts, we are assisted by a local newspaper report, published the day before their arrival, in which Hildreth explained that the station's overall task was to introduce superior plants to the people and communities of the High Plains.[1]

The mission of the Cheyenne Horticultural Field Station fit into the purview of the New Deal, although the station's founding preceded the New Deal. Actually, the station came into existence not through any systematic analysis of regional horticultural needs but by a single act of pork barreling sponsored by Wyoming senator Warren on behalf of his close friend and business associate, George E. Brimmer (1879–1956). Known as a skilled behind-the-scenes attorney who avoided general practice and rarely went to court, Brimmer was a lifelong enthusiast for community beautification.[2]

"Pork barreling," a phrase adapted from the medieval practice of preserving salted pork in wood barrels, then distributing slabs to serfs or tenants, entered the late-nineteenth-century American colloquial vocabulary to describe the use of general taxpayer funds for the benefit of particular areas or individuals on the basis of political favoritism. No one has ever surpassed the Honorable Francis E. Warren (1844–1929), U.S. senator from Wyoming for thirty-seven years and chair of the Senate Appropriations Committee during the 1920s, as master of the federal pork barrel.

Despite the circumstances of its establishment, the Cheyenne station blossomed into the leading horticultural institution serving the High Plains from 1930 until 1974. As an after-the-fact rationale for the station, it has been stated frequently and publicly that any plant surviving Cheyenne's climate should be able to thrive anywhere on the High Plains. Among the pertinent statistics: Cheyenne's elevation is 6,200 feet, with a correspondingly short growing season; Cheyenne holds the dubious honor of having the most frequent occurrence of hailstorms of any American city, averaging ten per year; Cheyenne is rated the fourth-windiest city in the United States, with an average wind speed of thirteen miles per hour; and Cheyenne temperature extremes, year-round, can range by fifty and even sixty degrees within twelve hours or less. In addition, precipitation is barely fourteen inches per year; while the soil, which is quite porous, has a low water-holding capacity. If hail is the greatest threat to horticulture in summer, the desiccating combination of low temperature, low humidity, limited snow cover, and wind is the greatest threat in winter.[3]

The story of the Cheyenne Horticultural Field Station begins in 1927. In late September of that year, Senator Warren and friend Brimmer paid an unannounced visit to the United States Department of Agriculture (USDA) Northern Great Plains Field Station in Mandan, North Dakota. Brimmer had been flooding Warren with background information; Warren had already met with Secretary of Agriculture William M. Jardine and Forest Service director William B. Greeley in Washington, D.C. Warren had initially opted for a federal nursery near Cheyenne to support forest restoration along upper Crow Creek and Pole Mountain, an area between Cheyenne and Laramie that happened to be leased to or owned by the Warren Livestock Company. After the senator concluded in the summer of 1927 that any new USDA nursery would likely be located far from Cheyenne, he provided Brimmer with the opening to push aggressively for a new, federally funded Central High Plains Field Station in Cheyenne, comparable to the Northern Great Plains Field Station in Mandan and the Southern Great Plains Field Station in Woodward, Oklahoma.[4]

As early as April 1927, A. L. Nelson (no relation to Aven Nelson), superintendent of the Cheyenne Experiment Station, which had moved to Archer nine miles east of town and was limited to dry-land farming, had gotten a hint that Brimmer and Warren were concocting something; he conveyed his concern to John M. Stephens, senior agriculturalist and head of dry-land agriculture at Mandan. Stephens replied that Nelson should well be able to give reasons why another field station near Cheyenne was not needed: "It seems to me that the Archer station, with the line of work you have going there, should be sufficient for the surrounding territory for some time to come. It would be better for your enthusiastic friends [Brimmer and his Cheyenne business compatriots] to assist you in getting adequate money to build up your own plant rather than start a new one."[5]

Since Archer was a cooperative project of the USDA and the University of Wyoming, agricultural experiment station director W. T. Quayle objected "very strongly" to the idea of a new station. He conveyed the university's opposition to Warren directly during a meeting, sometime in August 1927, in the senator's Cheyenne office and also attended by Stephens (at the request of Secretary Jardine), Brimmer, and local officials. At that meeting, Warren promised that, while he wanted a horticultural experiment station in Cheyenne, he would only support something that could be reasonably justified. From his perspective, the station first needed the approval of both Wyoming and USDA officials. "As matters now stand," Stephens reported to USDA headquarters, "it is a battle between the Cheyenne and the State people, and I hardly think we will have anything to worry about concerning it, as least for some time."[6]

Yet, at that same meeting and after Warren's promise, Stephens received more questions along a rather suspicious line: if matters could be arranged to satisfy the university regarding Archer, then what would a new station cost, and how much land and what water rights would be needed? Based on his experience at Mandan, Stephens came up with the figure of $100,000 for construction plus the contribution of at least two sections (1,280 acres). Warren, Brimmer, and local officials then proceeded to show Stephens their intended site, approximately 2,000 acres of municipal property held for park expansion combined with another 240 acres of private land, located approximately six miles northwest of Cheyenne. Stephens received assurance that the city would purchase the private land, then deed both properties and water rights to the USDA at no cost.[7] Stephens apparently did not know that Warren had already obtained Secretary Jardine's support for special legislation to establish and appropriate monies to build and operate the new station.[8] And so it was, with preliminaries completed, that Brimmer and Warren inspected the Mandan station the next month, September 1927.

By then, midlevel officials at the USDA had become aware of the pressure exerted by Cheyenne "business interests," namely, by Brimmer's barrage of letters and telegrams to Senator Warren and Secretary Jardine as well as to President Arthur G. Crane at the University of Wyoming. By December 1927, the USDA officials accepted the inevitability of a new station and scrambled to prevent existing and proposed appropriations from being deflected on behalf of Cheyenne. "It should be clearly and definitely understood," wrote E. C. Chilcott, longtime head of the Office of Dry Land Agriculture, "that no one officially connected with the USDA has in any way stimulated or encouraged these proposed increases in appropriations [for Cheyenne]." Chilcott feared most of all the prospect of cutting short research at Archer on crop rotation and tillage methods, which had been conducted successfully for fourteen seasons and required several more years to yield useful results.[9]

Since Brimmer wanted community beautification, Warren's bill to establish the Cheyenne station provided exclusively for horticultural experiment and demonstration work and omitted dry-land crops, thus preserving research at Archer:

> That the Secretary of Agriculture be, and he is hereby, authorized and directed to cause such shade, ornamental, fruit, and shelter-belt trees, shrubs, vines, and vegetables as are adapted to the conditions and needs of the semiarid or dry-land regions of the United States, to be propagated at an experiment station of the Department of Agriculture to be established at or near Cheyenne, Wyoming, and seedlings and cuttings and seeds of such trees, shrubs, vines, and vegetables to be distributed free of charge under such regulations as he may prescribe for experimental and demonstration purposes within the semiarid or dry-land regions of the United States.

The bill, signed by President Calvin Coolidge on March 19, 1928, authorized an appropriation of $100,000. Both Secretary Jardine and President Crane strongly endorsed the bill during congressional hearings.[10]

Within a month of the bill's passage, Chilcott from Washington and Stephens from Mandan were dispatched to Cheyenne. Their major challenge was to accommodate the senator, who kept his hands in the developing details until his death on November 24, 1929, and Brimmer, who, according to his friend Warren, had declined to stand by and let the experts do their work.[11]

And work there was, in securing the land, fencing the property, installing an irrigation system, and building station facilities. Brimmer had skillfully obscured the fact that 44 acres of the most strategic private land had been omitted from the proposed lease agreement. As soon as Chilcott discovered that, he got the city to agree to institute condemnation proceedings; after that, on June 6, 1928, the agreement was executed, with the city leasing to the USDA 2,135 acres for ninety-nine years at the rate of one dollar per year, renewable not beyond June 30, 2126, and granting 450 acre-feet per year of wastewater from the municipal filtration plant on nearby Roundtop at no cost to the station.[12]

During the first six months of site development and building construction, Chilcott and other USDA specialists made regular trips to Cheyenne, while Stephens remained at the site virtually full-time to oversee construction. Saco Rienk DeBoer, the Denver landscape architect who had developed Cheyenne's tree plan, prepared the layout for the station grounds; William Dubois, a preeminent Cheyenne architect, designed the buildings. Contractors completed staff residences, mess hall, bunkhouse, office, a combination seed, packing, and implement shed, two barns, storage cellar and work room, and garages during the first two years and constructed two greenhouses (complete with wire mesh screening to protect against hail) and a lathe house (used to acclimatize plants) during the third year. In addition, they installed over twelve miles of fences and, under the auspices of the Bureau of Public Roads, began installation of an irrigation system. A series of headgates and ditches leading from Roundtop reservoir was placed in time for the 1930 growing season. The completed irrigation system included roughly 2.5 miles of concrete-lined ditches built by the Civilian Conservation Corps.[13]

George Brimmer insisted that experimental work begin in spring 1930, even if that meant postponing some construction; he sought to intimidate the USDA into making personnel and budgetary adjustments toward that end. During the week in which his friend Warren died, Brimmer visited the station, reviewed financial statements, and concluded—incorrectly—that

Cheyenne Horticultural Field Station, early 1930s. Courtesy, USDA–Agricultural Research Service.

Cheyenne Horticultural Field Station, early 1930s. Courtesy, USDA–Agricultural Research Service.

Cheyenne Horticultural Field Station, late 1930s. Courtesy, USDA–Agricultural Research Service.

experimental work was scheduled to start in the spring of 1931. His misunderstanding sufficiently alarmed Chilcott to send Stephens to Cheyenne at once. After two lengthy meetings with Brimmer, Stephens reported back that "we will be laying ourselves open to severe criticism if we do not start the field work rather vigorously in the spring [of 1930], and so far as I know you are planning to do so."[14]

As instigator of the project and the person responsible for obtaining land and water from the municipality, Brimmer had become the target for burghers who criticized the seemingly slow pace of work at the station. In particular, he received pressure from members of the Cheyenne Industrial Club, precursor to the Chamber of Commerce, who expected the station to contribute more to the local economy than to community beautification. And that gave Brimmer a new argument for expediting work at the station: it would attract tourists. Cheyenne's location as a transportation hub provided travelers with the rare opportunity to view experiment station work firsthand and, at the same time, helped fill the city's restaurants and hotels.[15]

In the spring of 1929, Captain Robert Wilson, an arboriculturist who came from Mandan to serve as the station's first superintendent, oversaw planting of the first shelterbelt. Seedlings came from Mandan, the Bessey

Lathe house in which tender plants were started, Cheyenne Horticultural Field Station. Courtesy, USDA–Agricultural Research Service.

Nursery, and the newer Forest Service nursery near Monument, Colorado. While construction of physical facilities took place, workers did prepare the grounds for orchards, vegetables, and ornamentals.[16]

As the order of work shifted from construction to planting, Dr. William A. Taylor, director of the Bureau of Plant Industry, transferred the station from the Office of Dry Land Agriculture to the Office of Horticultural Crops and Diseases, except that shelterbelt and related agronomic activities remained with Dry Land Agriculture. Brimmer had been advised that this would happen, and he could find no objection.[17]

Brimmer had not been told, however, that the administrative realignment provided the opportunity for hiring a professional horticulturist to supervise the station. USDA officials knew that Brimmer had developed a controlling relationship with Wilson. With the appointment of Dr. Hildreth and the departure of Captain Wilson, Brimmer's interference in station management effectively ended.

A native of West Virginia, Aubrey Clare Hildreth (1893–1975) had graduated with a degree in agriculture (horticulture) from West Virginia State University in 1917. After a stint as county agent, then a tour of duty overseas

Ditches showing control gate open, laterals closed, Cheyenne Horticultural Field Station. Courtesy, USDA–Agricultural Research Service.

in the military, Hildreth returned as research fellow and instructor at the University of Minnesota, wrote his thesis on "Determinations of Hardiness in Apple Varieties and the Relation of Some Factors to Cold Resistance," and earned a Ph.D. in 1926. Before taking the Wyoming position, Hildreth was employed at the Washington Agricultural Experiment Station, working on dry-land plants, and at the Maine Agricultural Experiment Station, on improving the native blueberry (*Vaccinium angustifolium* Ait.).[18]

Clearly, the USDA had found an experienced professional horticulturist to develop the research program at Cheyenne. To begin with, Hildreth organized the station's research into four divisions of plant materials (fruits, vegetables, windbreaks, ornamentals) and two additional divisions that cut across plant lines: plant physiology and plant breeding. For each of the six divisions, Hildreth requested a specialist and one or more assistants. Because of smaller than anticipated and continually fluctuating budgets, he was able to staff fully only the divisions of vegetables, windbreaks, and plant breeding—and those not all the time. Hildreth himself headed the plant physiology division and directly oversaw work in fruits and ornamentals.[19]

At first, Hildreth was the only Ph.D at the station; however, he encouraged his professional staff to pursue graduate studies. Myron F. Babb, the vegetable specialist, earned a Ph.D. at the University of Minnesota; James E. Krause, Babb's assistant, earned a Ph.D. at Cornell University; Harold F. Engstrom, windbreak specialist, took graduate study at the University of Minnesota; Engstrom's successor, John L. Emerson, attended the University of California, Berkeley; and LeRoy Powers, geneticist and plant breeder, who joined the staff in 1935, earned a Ph.D. at the University of Minnesota.[20]

The station's support staff varied in number depending on budgets, at times consisting of ten, not counting seasonal laborers and Civilian Conservation Corps workers: plant propagator, farm foreman, herdsman, head teamster, tractor man, truck driver, greenhouse helper, mess cook/caretaker, janitor, and clerk. In the matter of personnel management, Hildreth meant to hire competent people, enable them to do good work, and supervise them informally. He was "so very casual," according to a former seasonal worker, that his subordinates did not realize when he was teaching them.[21]

Hildreth created, and the station followed, roughly the same work plan for each of the four plant materials divisions: to collect all varieties that might succeed on the High Plains; to determine the adaptability of those varieties through field tests at the station and at cooperator sites; to select and release to the public those varieties considered worth cultivating in the region; to breed new varieties in case no suitable varieties existed to meet certain conditions; to develop the most suitable cultivation practices under dry-land and irrigated conditions; and, concurrent with the plant materials work, to investigate the genetics of the plants and the physiology of their adaptation to the conditions of cold, drought, alkaline soil, dry atmosphere, and high light intensity characteristic of the region.[22]

Before leaving the station scene, Brimmer had recommended that, to save the salaries and expenses of two botanists to collect native seeds and plants, contact be made with Aven Nelson, who had already "collected specimens of approximately 100,000 varieties of trees, shrubs, flowers and grasses in the State of Wyoming."[23] Although the purpose of a university herbarium is different than that of a horticultural experiment station, Hildreth and Nelson ended up doing considerable collecting together. In 1932, for example, they visited Granite Canyon, Dale Creek, and Crow Creek in southeastern Wyoming and the Sandhills near Ellsworth, Nebraska. For two months that summer, Hildreth contracted with Nelson—by then a sprightly seventy-three—which resulted in a collection of over 200 different species, all eventually planted at the station. All professional staff at the station consistently sought out native varieties for possible cultivation, either alone or crossbred with non-native varieties.[24]

The station secured seeds and plants from the world over, through various USDA divisions; directly from commercial seed houses and nurseries, experiment stations, botanic gardens, and arboreta; and from individuals living within and beyond the boundary of the High Plains. To be sure, many plant materials came from botanists and horticulturists through the nearby land-grant colleges. In addition, the station accession log reveals that by the 1930s, at least a dozen nurseries along the Front Range were contributing plant materials. Most notable were W. W. Wilmore Nurseries in Wheatridge, Rockmont Nursery in Boulder, Loveland Nursery and Orchard, and Northern Colorado Nursery in Fort Collins—all in Colorado. Outside the region, but within the High Plains states, were Hansen Nursery in Brookings and Gurney in Yankton, South Dakota; Plumfield Nurseries in Fremont and Marshall Nursery in Arlington, Nebraska; and Willis Nursery in Ottawa near Topeka, Kansas.[25]

With the grounds prepared for cultivation in 1929, actual planting of fruit trees began for the dry-land orchard in 1930 and for the irrigated orchard in 1931, altogether about 100 acres located near the east entrance to the station grounds. In his summary of thirty years of research, written after he left Cheyenne, Hildreth noted that more than 2,000 fruit varieties had been tested through the station. Again, understand that by "variety," the horticulturist means a "sport" or plant that exhibits some striking variation from its parent stock, not a "variety" in the way a botanist might use the term "subspecies." In the case of the Wealthy apple (*Malus pumila* Mill. cv. "Wealthy"), still popular in Hildreth's day, researchers created a number of specimens for particular characteristics such as cold resistance and assigned each a unique number but did not name them as a separate "sub-variety." Still today, even from the best commercial nurseries, one is never sure, for example, whether a specimen of *Malus pumila* Mill. cv. "Haralson" is truly the cultivar by that name or a further subdivision one might call a "sub-cultivar." For the varied and difficult growing conditions on the High Plains, these seemingly minor nomenclatural distinctions can make the difference between thriving and surviving and sometimes between survival and death.[26]

Fruit research at Cheyenne took place on the generally accepted premise that large commercial orchards were impractical for the High Plains but that fruit could be grown for home use and local markets. Hildreth's plan called for planting at least three specimens with similar characteristics and, when loss occurred, to replant during the next growing season. In practice, more

Irrigated orchard with alfalfa between tree rows, Cheyenne Horticultural Field Station. Courtesy, USDA–Agricultural Research Service.

or fewer trees of a single variety were planted; generally, fewer than three of a variety survived. Hundreds of apples and pears were planted through 1945, with plums and cherries planted through 1949. Only an occasional fruit tree remains because the orchards were uprooted and the land cleared to make way for the new emphasis on grassland research.

During the station's horticultural heyday, however, varieties placed in the orchards generally resulted from grafting onto rootstocks. In the case of apples and crab apples, the preferred rootstocks were hardy crab apple cultivars such as *Malus ioensis* (A. Wood) Britt. cv. 'Dolgo,' then *Prunus americana* Marsh. for plums, *Prunus cerasifera* J.F. Ehrh. or *Prunus Mahaleb* L. for sour cherries, and rootstocks unknown for pears. To ensure pollination, the station brought in several hives of honey bees to supplement native bees and other pollinating insects.

Initially, researchers planted fruit trees fifteen feet apart within rows and thirty feet between rows. In 1942 every other tree in the rows was removed to make tractor cultivation easier and, incidentally, to conserve more moisture. The dry-land orchard received no irrigation, not even in the driest years; the irrigated orchard received two heavy applications, in midsummer and late fall. No fertilizers were used on either orchard, but both were sprayed against coddling moth and canker worm. Fire blight, the pervasive bacterium (Micrococcus amylovorus) that attacks especially apples and pears, rapidly turning their leaves brown, was controlled by pruning infected trees. In fact, regular and severe pruning occurred on all fruit trees to record better bloom and yield information. Any produce went to station employees and to local charities for distribution to the needy.[27]

Of the more than 400 varieties of apples and crab apples tested, around 150 failed. Among those that survived, researchers identified thirteen apples and six crab apples hardy and adaptable enough to recommend for home and local market use.[28] Of the apple varieties, Cortland, Haralson, Red Duchess, Redant, and Wealthy remain available commercially, although some only from specialty nurseries. Of the crab apples, Columbia and Dolgo remain available, with Dolgo perhaps the most suitable to the High Plains.

Although sweet cherries do not thrive on the High Plains, Hildreth and staff undoubtedly knew about the early orchardists in northern Colorado experimenting with sour or pie cherries (Prunus cerasus L.). Of the twenty-one horticultural varieties that grew at the station, researchers recommended four that remain available today: Early Richmond, English Morello, Meteor, and Montmorency. In addition, building on the work of Niels Hansen, Hildreth crossed the western sand cherry (Prunus pumila L. var. besseyi (Bailey) Gl.) with varieties of the Nanking cherry (Prunus tomentosa Thumb.) in the continuing, although unsuccessful, search for a commercially viable cherry adapted to dry-land cultivation.[29]

Of nearly 200 additional varieties of Prunus tested, the hardiest were varieties of native plum (Prunus americana Marsh.) crossed with varieties of the Japanese plum (Prunus japonica Thunb.). Among those hybrids, four can still be purchased through specialty nurseries: Fiebing, La Crescent, Pembina, and Tecumseh. The station received for testing more than 50 varieties of pears (Pyrus L), like plums, part of the rose family. Some varieties came from northern China; fewer than half survived at Cheyenne, and none could be recommended for home and garden on the High Plains.[30]

Compared to the orchard fruits, the small fruits—currant, gooseberry, raspberry, and strawberry—showed far better results. Among varieties tested, Red Lake currant as well as Pixwell and Red Jacket gooseberry can still be purchased commercially. Most notable, breeding work on raspberries and strawberries eventually resulted in the release of four new varieties still available from specialized commercial nurseries: Pathfinder and Trailblazer raspberries and Fort Laramie and Ogallala strawberries.

When Hildreth and Powers began their research in 1932, no commercial variety of strawberry existed that could withstand the climate of the High Plains. During a period of nearly fifteen years, but especially from 1935 until 1945, Powers collected more than 42,000 specimens of the "native Rocky Mountain strawberry" (*Fragaria virginiana* Duchesne) from 1,100-plus locations in Wyoming and Colorado, as well as in Montana, New Mexico, and Utah. None of these specimens was suitable for commercial use, mainly because of their small fruit but also as a result of their uneven qualities. By pollinating these hardy natives with three commercial cultivars—Gem, Dorsett, and Fairfax—Powers did manage to create flavorful, attractive, and hardy varieties suitable for the High Plains.

Powers conducted similar experiments with raspberries, using the native *Rubus idaeus* L. to inculcate hardiness and commercial varieties to breed for size and other qualities. Raspberries alone occupied nearly ten acres at the station. Given the high cost of such research, the Cheyenne creations went primarily to provide commercial nurseries with hardy parental material.[31]

The greater part of station research took place with vegetables, with more than 8,000 varieties tested at Cheyenne and on cooperating farms, ranches, and experiment stations throughout the High Plains. Because of the relatively large number of varieties in each species, the likelihood of finding suitable varieties was the greatest. Concerning annuals, researchers sought out varieties that could survive the climate extremes; for biennials and perennials, they looked for varieties that could survive climate extremes aboveground during growing seasons and belowground through dormancy. Varieties attracting the greatest interest were those that matured with little or no irrigation—for example, asparagus, beets, cabbage, carrots, cucumbers, onions, orach, parsnips, pumpkins, rhubarb, rutabagas, spinach, squash, Swiss chard, and tomatoes. In addition, the hardy perennials such as asparagus, chives, and rhubarb matured before the last frosts of spring, while varieties such as beets, cabbage, carrots, cauliflower, kale, parsnips, rutabagas, and turnips prolonged the season beyond the first frosts of fall. Furthermore, several root vegetables as well as potatoes and tomatoes seemed to recover well from the ravages of the inevitable hail.[32]

Under pressure to start experiments, with the station under construction and no horticulturist on staff, Captain Wilson had secured enough vegetable seeds and seedlings from Mandan to plant the first five-acre plot. He had not realized, however, that the selected site contained a heavy infestation of whitetop (*Cardaria draba* (L.) Desv.), a noxious weed difficult to eradicate because of its spreading underground roots. As a result, combined with an early hard frost, the first-year harvest was of no practical value.[33]

During the 1931 growing season, Myron F. Babb oversaw plantings of more than 650 varieties of vegetables, and visiting horticulturists planted another 133 varieties of beans and 35 varieties of Jerusalem artichoke (*Helianthus tuberosus* L.), a species of sunflower that produces potato-like tubers that taste somewhat like artichoke. By means of parallel plantings in 1932, Babb illustrated that vegetables under irrigation yielded four to eight times more produce than dry-land yields and were more valuable economically, even after taking into account the higher costs of irrigated cultivation.[34]

In searching for ways to lengthen growing seasons, Babb conducted research on how best to prepare young plants as they were moved from protected environments such as hotbeds, cold frames, and greenhouses into unprotected gardens and fields. His doctoral dissertation, completed while he worked at the station, compared the traditional method of "hardening off" with the newer method of "forcing" plants. By "hardening off," he meant slowly weaning young plants from artificially high moisture and warm temperature conditions before transplanting them into gardens or fields. By "forcing," he meant artificially supplying plants with supplementing nutrients, maintaining optimal temperature and moisture conditions until they could be transplanted safely. Babb concluded that most vegetables respond better to forcing than to hardening off, although horticulturists today tend to combine both methods, especially when trying to grow warm-weather, long-season vegetables on the High Plains.[35]

Because of the overwhelming popularity of the tomato, Babb maintained testing of approximately 400 varieties over a period of several years. He recognized, however, that no single variety could do well under the varied conditions of all sections of the High Plains. Since the extent of his experimenting meant not every variety could be observed and tested in a single season, Babb did grow the same well-known, commercial varieties each year as standards by which to judge the performance of all varieties. Among them, the Bison, Bonny Best, Earliana, and Marglobe varieties remain available from specialty houses. Generally speaking, most commercial varieties did not consistently set fruit either because of the penetrating summer heat or because the time needed for their maturity was too long.

To help overcome those obstacles, Babb obtained, through the USDA Division of Foreign Plant Introductions, a number of varieties that matured considerably earlier, although their fruit size and yields were less than the standard varieties. H. O. Werner at the agricultural experiment substation in North Platte worked jointly with Babb, most notably on the Danmark tomato variety obtained from a seed company in Grimstad, Norway. Danmark proved among the best varieties in setting fruit under extreme climatic conditions and the earliest variety to produce fruit acceptable for the commercial market. In addition, by importing relatively hardy varieties from places such as Siberia and pollinating them with the more common varieties, Babb and his successors managed to breed and release three new varieties—Alpine, Colorado Red, and Highlander—of which only the last is still available.[36]

Even if Babb had succeeded in creating a frost-resistant tomato, which he did not, obstacles remained to growing this plant successfully on the High Plains. At Hays, for example, with a longer growing season and lower elevation than Cheyenne, thirty-two varieties in 1938 succumbed to light frost shortly after they were moved into the fields. A second planting suffered severe hail damage, followed by an invasion of gray blister beetles that destroyed the first sets of blossoms. Field workers sought to hand-pick the beetles that appeared in waves throughout the remainder of the growing season. Despite decimation from hail, beetles, and grasshoppers, enough plants survived to study the effects of pruning and shade and to ascertain that the best variety for western Kansas was the Valiant, an indeterminate variety producing six- to eight-ounce-sized, globe-shaped fruit within roughly eighty days of transplanting. Indeterminate means a plant that grows continuously from all branches as long as it is alive and that is not self-pruning by ending growth in its flowering stalks.[37]

With the coming of World War II, station staff accelerated cooperative efforts to discover and recommend vegetables that did best, especially in Victory Gardens. Among lesser-known varieties, Babb had field tested orach (*Atriplex hortensis* L.) as a substitute for spinach. He found that orach, which tastes somewhat like spinach when cooked, was easily cultivated, exceptionally drought-resistant, and tolerant of alkaline soils. Orach starts seed stalk formation early, unlike spinach, which tends to "bolt," or run to seed, before the plant has grown enough to produce quality leaves; orach continues to put out large, tender leaves on the upper portion of the plant. Orach can grow up to six feet tall, providing a continual supply of greens throughout the growing season, and it produces an abundance of seeds that can easily be saved for the next season.[38]

Meanwhile, LeRoy Powers, the station geneticist, developed an early-maturing squash, *Curcubita pepo* cv. "Early Cheyenne pie pumpkin," which was just large enough to make one pie, although too small for the canning trade. He had selected early-maturing specimens of the commercial New England pie pumpkin, then inbred them by self-pollination for six generations until he had created a uniform fruit and plant type. In addition, as with the

Breeding tomatoes, Cheyenne Horticultural Field Station. Courtesy, USDA–Agricultural Research Service.

Portion of ten-acre breeding plot for early-maturing tomatoes, Cheyenne Horticultural Field Station. Courtesy, USDA–Agricultural Research Service.

Vegetable variety test plots, Cheyenne Horticultural Field Station. Courtesy, USDA–Agricultural Research Service.

Breeding early strain of pie pumpkin, Cheyenne Horticultural Field Station. Courtesy, USDA–Agricultural Research Service.

Frost covers for breeding melons, Cheyenne Horticultural Field Station. Courtesy, USDA–Agricultural Research Service.

Asparagus trial plots, Cheyenne Horticultural Field Station. Courtesy, USDA–Agricultural Research Service.

Beets: root, section, leaf, Cheyenne Horticultural Field Station. Courtesy, USDA–Agricultural Research Service.

fruits, station staff conducted studies on the effect of climate and storage on the vitamin content of vegetables; similar investigations took place on the carotene content of carrots.[39]

For the home gardener in Wyoming, Babb published what would become the standard reference manual, later expanded to include the entire Rocky Mountain region. Following a practical introduction to the location of gardens, seed selection, and cultivation, Babb set forth an annotated list of vegetables that, because they thrived at Cheyenne, were likely to do well throughout the High Plains. Babb classified this group as suitable for dry-land gardening (cultivar name given on varieties Babb tested, which are still commercially available):

> Perennial vegetables: asparagus Mary Washington, horseradish, garlic, rhubarb
>
> Cabbages: cabbage Golden Acre, cauliflower, broccoli
>
> Potherbs: collards, kale Dwarf Curled Siberian, spinach (substitute Swiss chard
>
> Lucullus for elevations above 6,000 feet)

Cauliflower plant ready to tie, Cheyenne Horticultural Field Station, September 1939. Courtesy, USDA–Agricultural Research Service.

Seedbeds, Cheyenne Horticultural Field Station. Courtesy, USDA–Agricultural Research Service.

Salads: lettuce Grand Rapids and Boston, parsley

Root crops: beet Detroit Dark Red, carrot, radish, rutabaga, turnip

Bush bean Pencil Pod Black Wax and peas

Tomato

Squashes: winter squash Golden Hubbard, summer squash, pumpkin Early Cheyenne, cucumber

Corn: Sweet corn Pickaninny (below 6,000 feet)

Onions: Spanish Early Grano (below 6,000 feet)

In the opinion of the Cheyenne researchers, the flavor and quality of vegetables grown at high altitude surpassed those grown at lower elevations and in warmer climates.[40]

While research on fruits and vegetables aimed to improve home and local-market gardening, research on ornamentals and trees intended to make home and community more attractive. In the wake of the Great Depression, many rural families clearly could no longer pick up and leave their homes. Instead, Hildreth and others observed that rural families decided to put down firmer roots by planting their surroundings, making their homes more conducive to comfortable living.[41]

Of over 1,300 varieties of woody ornamental perennials tested for adaptability through the station, Hildreth seemed most interested in about 57 varieties deemed most suitable for use as hedges: varieties of pea shrub (*Caragana arborescens* and *Caragana microphylla* from Siberia and northern China, respectively), cotoneaster (*Cotoneaster acutifolius*, *Cotoneaster lucidus*, and *Cotoneaster racemiflorus soongoricus*, also from Asia), cinquefoil (*Potentilla parvifolia* cv."Farreri"), chokecherry (*Prunus virginiana var. demissa*), buckthorn (*Rhamnus cathartica* [common] and *Rhamnus davurica* from Japan and Korea), and lilac (*Syringa* x *persica*). In addition, Hildreth created a hardy variety of privet, *Ligustrum vulgare* cv. "Cheyenne," still commercially available. Although privet was the most common American hedge plant at the time, Hildreth's release appeared to be the first drought-resistant and alkaline-tolerant variety.[42]

Seeking to promote the ornamental use of hardy plants, Hildreth collected over 700 different varieties of herbaceous (non-woody) perennials. In addition to their ability to withstand the combination of cold, wind, and low humidity, many of these perennials bloomed from early spring until late fall. Among them, the chrysanthemums attracted the most public attention. Indeed, the golden-centered, white-rayed ox-eye daisy (*Chrysanthemum leucanthemum* L.),

Flower beds, ornamental plantings, Cheyenne Horticultural Field Station, 1930s. Courtesy, USDA–Agricultural Research Service.

escaped from cultivation, is now commonly found along roads and ditches on the High Plains.

During the station's heyday, researchers conducted trials of more than 1,000 varieties of chrysanthemums, covering over two acres at the station; at the same time, they oversaw cooperative tests at Denver city parks, at experiment substations in Colby and Hays, Kansas, and at commercial nurseries nationwide. Hildreth and his successor, Gene S. Howard, would release more than 45 varieties of the Cheyenne Hardy Mum—all bred to bloom early for short growing seasons and to survive harsh winters. Again, the greatest value of these new releases was as parental material for commercial nurseries, which themselves could not absorb the research and testing costs.[43]

Keeping in mind that protection against wind is prerequisite for any horticulture on the High Plains and that the cultivation methods for starting trees on the High Plains were more similar to horticulture than to forestry, it made sense to have shelterbelts as a major topic of research at the Cheyenne Horticultural Field Station. To be sure, work on trees and shrubs had been taking place at Hays since 1901 and at Akron since 1909. In addition, the USDA Bureau of Plant Industry had been testing trees for adaptability at its stations in North Platte, Nebraska, and Ardmore, South Dakota (on the state line just north of Crawford, Nebraska), since 1907 and 1916, respectively. Supported in part by Clarke-McNary appropriations, the Forest Service had been growing seedlings for both mountains and plains in its Halsey, Nebraska, and Monument, Colorado, nurseries.

In 1930 the USDA transferred both shelterbelt stock and its agreements with cooperators on the High Plains from Mandan to Cheyenne. That move accelerated the cultivation and trials of both native and foreign trees and shrubs at Cheyenne and led to the establishment of an arboretum, a plot set aside for the cultivation and exhibition of trees and shrubs.[44]

With the cessation of shelterbelt operations at Mandan, its chief arboriculturist, John L. Emerson, transferred to Cheyenne where he introduced the technique of inter-planting seedlings with fast-growing annuals—sunflowers, sorghums, field corn, and hemp. In addition to providing summer protection against desiccating winds, these temporary windbreaks tended to collect snow during the winter, thereby providing added moisture during the early spring.[45]

During 1935 alone, Emerson oversaw the propagation of more than 1.3 million trees and shrubs, representing 128 different species and 688 varieties, most unknown to the region's commercial nurseries. He selected approximately 200 of those varieties for further testing and observed that, for optimal growth and broadest protection, shelterbelt trees and shrubs needed to be planted farther apart than had been the custom. Among deciduous trees, the Siberian elm (*Ulmus pumilla* L.) proved the outstanding species for dryland shelterbelts, while among conifers, the ponderosa pine, red cedar, and Colorado blue spruce remained the favorites.[46]

Perhaps more so with trees and shrubs than with the other plant materials divisions, the station relied on cooperators to ascertain whether promising plantings could really thrive throughout the region. Emerson oversaw experimental shelterbelt plantings at nearly 500 locations throughout the High Plains, except in western Nebraska, where much work had already been done. The station supplied to cooperators free of charge the nursery stock, landscape plans, and instructions for planting and maintenance. The cooperators, in turn, agreed to follow the instructions, pay for transportation of nursery stock from Cheyenne, and provide the land, necessary fencing, labor, and equipment for planting and maintenance. County agents often assisted station staff in the selection of cooperators.[47]

In general, cooperators planted shelterbelts around their homes and outbuildings. Station staff planned to inspect the shelterbelts regularly, but a combination of budget cuts, material shortages, and staff absences during wartime meant such visits did not occur. When Gene Howard visited some of these sites in the late 1950s, he discovered that he had been the first station visitor in fifteen years or more.[48]

Despite the difficulties of wartime and its aftermath, horticultural work did continue at the Cheyenne station. Encompassing all four divisions of plant materials, LeRoy Powers expanded his work on certain methods of genetic research begun at the University of Minnesota while Hildreth and others investigated ways to cope with iron deficiency, or chlorosis. That condition affects plants not acclimatized to highly alkaline soils and manifests itself when leaves become pale green, yellow, or brown, causing death in the most severely affected plants. Clearly, the best approach was to avoid species and varieties susceptible to chlorosis. The next-best approach, proven by trials overseen by Hildreth at the Denver parks, was the application of iron salts (ferric ammonium citrate) either by spraying leaves, getting into root zones, or injecting into tree trunks.[49]

The distinction between divisions of plant materials was not always hard and fast. Some ornamental trees and shrubs made good shelterbelts; commonly used shelterbelt varieties made good shade trees; and fruit trees, small fruits, and even some vegetable varieties worked as part of shelterbelts. Regardless of plant divisions, researchers at Cheyenne aimed first and foremost to demonstrate to the public the horticultural possibilities on the High Plains, but never at the expense of a better understanding of the overall nature of plants.

Fatigued by continuing budgetary deficiencies, Hildreth took retirement under the federal system in 1959 and accepted the invitation to take charge of the Denver Botanic Gardens, recently relocated to its present site. Not long afterward, the USDA made several attempts to close the station, for both financial and programmatic reasons. Each time, the Wyoming congressional delegation successfully prevented closure.

Lawrence A. Schaal, a specialist on potatoes, succeeded Hildreth as station superintendent; in 1964 he was followed by Gene Howard (1915–1999). An Oklahoma native and graduate of Southwestern State Teachers College at Weatherford (now Southwestern Oklahoma State University), Howard had moved to Cheyenne to work at the Frontier Refinery; in 1947 he was hired as field foreman by Hildreth, who saw in him a keen observer. Howard used data collected through the station as the basis for his master's thesis on shelterbelt trees and shrubs grown under dry-land culture on the High Plains, completed in 1961 at the University of Wyoming. Then, after twenty-five years at the station, Howard received the assignment of overseeing the transition from horticulture to grasslands research.

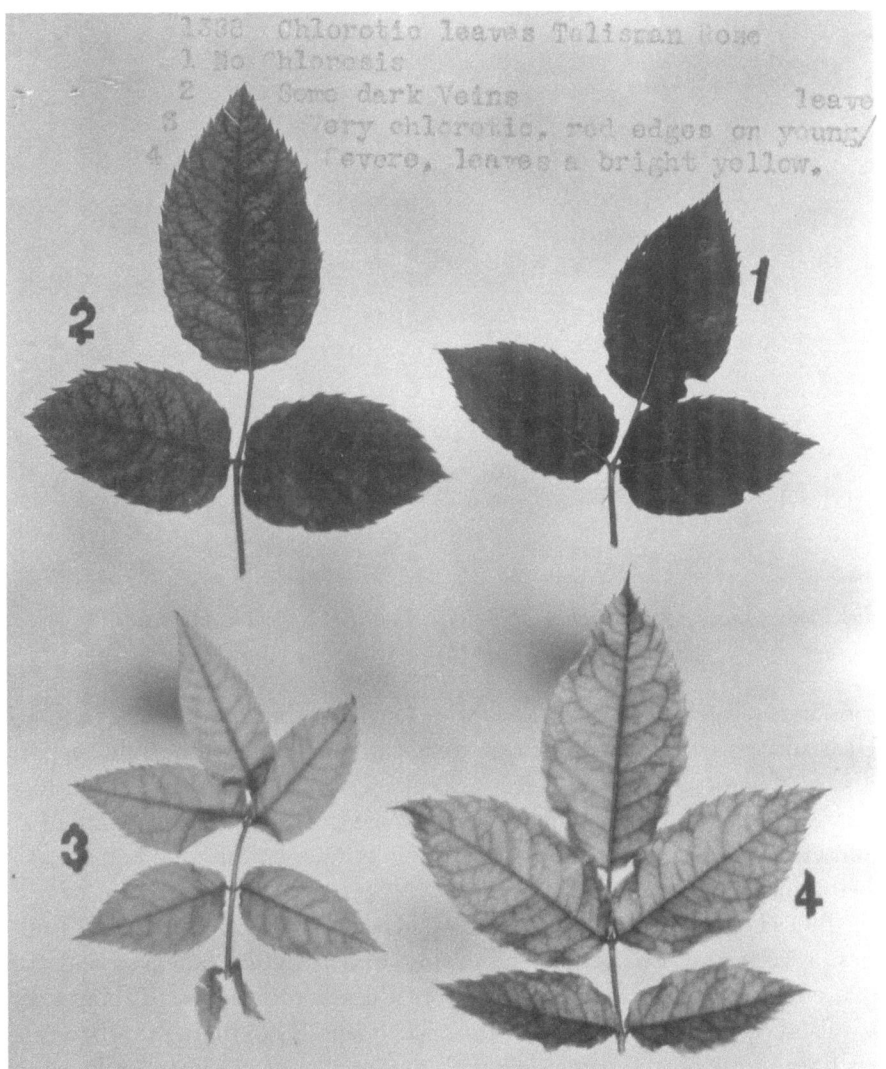

Chlorotic leaves, Talisman Rose: (1) no chlorosis, (2) some dark veins, (3) very chlorotic red edges on young leaves, (4) severe chlorosis, Cheyenne Horticultural Field Station. Courtesy, USDA–Agricultural Research Service.

Although the sequence of actions leading to the transition is obscure, certain aspects are clear: for some time, the Bureau of Plant Industry had wanted to close the station, its lack of enthusiasm perhaps stemming from the political motive underlying the station's establishment. To preserve the station,

High Plains Grasslands Research Station, 2007. Courtesy, Ronald K. Hansen.

grasslands researchers within the USDA sought to shift the station's mission. Their success can be attributed in great part to the active outside support of U.S. senator Gale W. McGee, a former University of Wyoming history professor. Both McGee and his Wyoming colleague in the Senate, Clifford P. Hansen, a rancher, viewed the new mission as an opportunity to better serve their rangeland constituents.[50]

By 1974, all horticultural plantings had been moved to other USDA facilities or destroyed, except for landscaping around station buildings and the arboretum. According to a 1974 inventory, the arboretum covered about forty acres and contained 851 tree and shrub specimens. Since then, the arboretum has been pretty much abandoned, with the number of specimens reduced to less than half. Occasional harrowing by station staff has helped control weeds, and the voluntary work of one station employee in particular has ensured the survival of several specimens. Recent station research leaders, headquartered in Fort Collins, have allowed commercial horticulturists to collect seeds and take occasional cuttings from the arboretum.[51]

As a result of feature stories about the former Cheyenne Horticultural Field Station in regional newspapers, three former Colorado State University horticulture students who had worked at the station organized a small group of individuals from the Cheyenne Botanic Gardens, commercial nurseries, and local, state, and federal agencies to provide basic maintenance at the arboretum. With support of the USDA through the Wyoming State Forestry Division, the group commissioned a master plan for the arboretum. The plan set forth a series of steps to preserve existing plants, make the arboretum accessible to the public, and allow it to serve once again as a source of useful horticultural knowledge to the increasingly urban population of the High Plains.

Whether the arboretum will be revitalized and whether the station's remaining horticultural facilities will be preserved as a historic site remains uncertain. What is known, however, is that even before the station's mission shifted to grasslands, horticulture on the High Plains had moved in a further, decidedly ornamental direction, promoted by local efforts and the emerging "green industry," in response to the unaesthetic aspects of urban development combined with the ever-present shortage of water.

Notes

1. "Wyoming Listed as Definitely for Roosevelt after His Visit," *Wyoming Eagle* (Cheyenne), October 13, 1936; "Portable Orchard at Horticultural Station," *Wyoming Eagle* (Cheyenne), October 10, 1936.

2. George Brimmer obituary, *Wyoming State Tribune* (Cheyenne), April 3, 1956.

3. Shane Smith, "The Hildreth/Howard Arboretum," White Paper (Cheyenne: Cheyenne Botanic Gardens, 2001), 1.

4. Brimmer to Francis E. Warren, Cheyenne, November 7, 1929, National Archives and Records Administration, Records of the Bureau of Plant Industry, Soils, and Agricultural Engineering, file 14979, box 679, National Archives, College Park, Maryland (hereafter cited as NARA; all NARA entries are file 14979, box 679, stack 170, row 26, compartment 21, shelf 7), refers to 1927 visit; Warren to Brimmer, Washington, D.C., May 5, 16, 19, 31 and June 6, 22, 1927, subgroup 1, series 1, carton 21, Francis E. Warren Papers, University of Wyoming, Laramie (hereafter cited as Warren Papers).

5. J. M. Stephens to A. L. Nelson, Mandan, May 5, 1927, NARA.

6. J. M. Stephens to E. C. Chilcott, Mandan, August 30, 1927, p. 2, NARA.

7. Ibid.

8. Warren to Brimmer, Washington, D.C., June 28, 1927, subgroup 1, series 1, carton 21, Warren Papers.

9. Memorandum, William A. Taylor, chief, Bureau of Plant Industry, Washington, D.C., to F. M. Russell, assistant to the secretary of agriculture, November 30, 1927, and memorandum, E. C. Chilcott to W. A. Taylor, December 5, 1927, both in NARA.

10. Act of March 19, 1928, *The Statutes at Large of the United States of America,* 70th Congress, 1st. sess., Ch. 228, 323; W. M. Jardine to Sen. Charles L. McNary, Washington, D.C., February 15, 1928, and Jardine to A. G. Crane, Washington, D.C., March 1, 1928, both in NARA.

11. Chilcott to Jardine, Cheyenne, April 21, 1928, and Taylor to Jardine, Washington, D.C., April 27, 1928, both in NARA; Warren to Brimmer, Washington, D.C.., June 13, 1928, subgroup 1, series 1, carton 21, Warren Papers.

12. Memorandum (not sent), Chilcott to Brimmer, May 19, 1928, and lease between City of Cheyenne and United States of America, June 1, 1928, both in NARA.

13. List of buildings constructed in 1929 in Taylor to G. F. Allen, Washington, D.C., October 19, 1929; greenhouse specifications, USDA Bureau of Plant Industry, September 17, 1929; irrigation plan, Carl Rohwer, associate irrigation engineer, to Chilcott, Fort Collins, August 27, 1929; irrigation cost estimate, Rohwer to Chilcott, September 9, 1929, all in NARA; Annual Report of the Cheyenne Great Plains Field Station (typescript), FY1930, 1, and FY1932, 3.

14. Brimmer to Taylor, Cheyenne, January 8, 1930, and Stephens to Chilcott, Cheyenne, November 29, 1929, both in NARA.

15. Brimmer to Warren, Cheyenne, November 7, 1929; Brimmer to Knowles A. Ryerson, Cheyenne, November 8, 1929; Brimmer to Taylor, Cheyenne, November 20, 1929; all in NARA.

16. "Annual Report," FY1929, 13.

17. Memorandum, Taylor to Chilcott, Washington, D.C., March 10, 1930, NARA.

18. Letter, Taylor to Robert Wilson, Washington, D.C., March 12, 1930, NARA; Fred C. Johnson, "New Director Appointed for Denver's Botanic Gardens," *The Green Thumb* (August 1959): 221.

19. Aubrey C. Hildreth, "The Cheyenne Horticultural Field Station," in William H. Alderman, ed., *Development of Horticulture on the Northern Great Plains* (St. Paul: Great Plains Region, American Society for Horticultural Science, 1962), 126–132.

20. Annual Report, FY1933, 4.

21. FY1931 Draft Budget, Cheyenne, August 13, 1929, NARA; James R. Feucht, conversation with author, May 16, 2005. Feucht was on staff summers of 1954 through 1956 assisting with squash- and strawberry-breeding trials.

22. Hildreth, "Field Station," 127.

23. Brimmer to E. C. Auchter, Cheyenne, June 11, 1930, NARA.

24. Annual Report, FY1932, 1, and FY1933, 4, 11.

25. Cheyenne Horticultural Field Station Accession Log Book (typescript), n.d., located at the High Plains Grassland Research Station, Cheyenne.

26. Hildreth, "Field Station," 127; "Cultivar," in *Hortus Third, a Concise Dictionary of Plants Cultivated in the United States and Canada* (New York: Macmillan, 1976), 344.

27. Gene S. Howard and G. B. Brown, "Twenty-Eight Years of Testing Tree-Fruit Varieties at the Cheyenne Horticultural Field Station, Cheyenne, Wyoming," *USDA*

Agricultural Research Service Crops Research Bulletin, series 34, no. 39 (October 1962): 4–5.

28. Ibid., 7–23; Gene S. Howard, "Recommended Horticultural Plants Generally Hardy and Adaptable in the Central Great Plains Region," USDA Agricultural Research Service B-770 (February 1982; repr. September 1999), 6 pp.

29. Report of the Central Great Plains Field Station, FY1932, 6, High Plains Grasslands Research Station, Cheyenne.

30. Howard and Brown, "Twenty-Eight Years," 24–40.

31. A. C. Hildreth and LeRoy Powers, "The Rocky Mountain Strawberry as a Source of Hardiness," *Proceedings of the American Society for Horticultural Science* 38 (1940): 410–412. In 1944, Powers identified the "native Rocky Mountain strawberry" as *Fragaria ovalis* (Lehm.) Rydb., now a synonym for *Fragaria virginiana* Duchesne *ssp. virginiana*. Since we are not sure where Powers did his collecting, and because some contemporary taxonomists believe the differences in *Fragaria virginiana* are too slight to merit division into subspecies, it is probably best to simply identify the "the Rocky Mountain wild strawberry" as *Fragaria virginiana* Duchesne. Powers, "Research Line Project Annual Report of Progress," K-1-3b (typescript, March 14, 1940) and K-1-5-1 (typescript, March 30, 1944), High Plains Grasslands Research Center, Cheyenne; Hildreth, "Field Station," 128.

32. Myron Francis Babb and W. L. Quayle, "Vegetable Culture and Varieties for Wyoming," *Wyoming Agricultural Experiment Station Bulletin* 250 (1942): 5. Seed samples from the Cheyenne Station are preserved at the National Center for Genetic Resources, formerly known as the National Seed Storage Laboratory, Fort Collins.

33. Report of the Central Great Plains Field Station, FY1929, 15, and FY1930, 3.

34. Ibid., FY1932, 23, 49, and FY1933, 59.

35. Ibid., FY1933, 49–54; see also Myron Francis Babb, "Residual Effects of Forcing and Hardening on the Morphology and Physiology of Vegetable Plants," *USDA Technical Bulletin* 760 (Washington, D.C., December 1941), 34 pp.

36. Myron Francis Babb and James E. Kraus, "Results of Tomato Variety Tests in the Great Plains Region," USDA Circular 533 (Washington, D.C., 1939): 1–11; James R. Feucht, "Field Station Recently Saved from Shut-Down," *Metro Horticulture* (October 1966): 1.

37. Hays Agricultural Experiment Substation Annual Report 1936 (typescript), 42–44; 1937, 39; 1938, 48–49 (Hays: Western Kansas Agricultural Research Station).

38. Myron Francis Babb and James E. Krause, "Orach, Its Culture and Use as a Greens Crop in the Great Plains Region," USDA Circular 526 (September 1939): 1.

39. LeRoy Powers, "Early Cheyenne Pie Pumpkin," USDA Circular 537 (November 1939): 1, 3; Aubrey C. Hildreth, "Horticulture on the Wyoming Plains," in William H. Alderman, ed., *Development of Horticulture on the Northern Great Plains* (St. Paul: Great Plains Region, American Society for Horticultural Science, 1962), 120.

40. Babb and Quayle, "Vegetable Culture," 39.

41. Report of the Central Great Plains Field Station, FY1934, 4.

42. A. C. Hildreth, "Ornamental Hedges for the Central Great Plains," *USDA Farmers' Bulletin* 2019 (1950): 13.

43. Report of the Cheyenne Horticultural Field Station, FY1936, 14; "Research Line Project Annual Report of Progress," K-5-2a-2 (March 18, 1940), and K-5-2-3 (March 4, 1942) (typescripts), High Plains Grasslands Research Center, Cheyenne; Gene S. Howard, "Herbaceous Perennials for the Central Great Plains," *USDA Agricultural Research Service Bulletin*, series 34, no. 71 (1965): 5–8, lists the most desirable perennials as determined by station field trials.

44. Memorandum, Auchter to Kellerman, Washington, D.C., December 10, 1930, NARA.

45. Report of the Central Great Plains Field Station, FY1932, 11–12.

46. Report of the Cheyenne Horticultural Field Station, FY1936, 35–36; Gene S. Howard and G. B. Brown, "Seven Species of Broadleaf Deciduous Trees for Windbreaks: Effect of Spacing Distance and Age on Their Survival and Growth at Cheyenne, Wyoming," *USDA Technical Bulletin* 1291 (1963): 3, 15.

47. Gene S. Howard, "A Study of Shelterbelt Tree and Shrub Species under Dryland Culture in the Central Great Plains Region" (M.S. thesis, University of Wyoming, 1961), 3, 13; George W. Boyd and Burton W. Marston, *The Wyoming Agricultural Extension Service and the People Who Made It, 1919–1964* (Laramie: University of Wyoming Publications, 1965), 44.

48. Howard, "Shelterbelt Trees," 15.

49. "Research Line Project Annual Report of Progress," K-5-2a-5 (March 18, 1940, and March 6, 1944) (typescripts).

50. Gerald E. Schuman, e-mail to author, November 28, 2005. Schuman was a longtime soil scientist at the Cheyenne station.

51. Gene S. Howard and Marilyn S. Samuel, "A List of Woody Ornamental Plants under Test at the Cheyenne Horticultural Field Station" (typescript, January 1974), 21 pp., located at High Plains Grasslands Research Center, Cheyenne. See also Scott T. Skogerboe, "Plants of the High Plains Arboretum" (January 1994), at http://www/botanic.org/ArboretumPlants.asp; accessed May 2008.

12

Horticulture and Community

The Lower Downtown Development District, affectionately known as LoDo, is a vital, attractive, and trendy part of contemporary Denver. Created in the late 1980s, the district has spurred a highly successful voluntary effort to balance economic growth and historic preservation. Amid the district's chic restaurants, sports bars, elegant boutiques, and loft apartments, the Rocky Mountain Seed Company building looks pretty much as it did when founded in 1920. Sales now depend mainly on wholesale and corporate buyers; until the summer of 2007, the home gardener or, more precisely, the weekend horticulturist could still purchase seeds by the packet off the racks or by weight out of the oak drawers and bins that lined the walls of this vestige of Denver's preeminence as the agricultural and horticultural center of the Rocky Mountain Empire. (Because of limited loading space downtown, new owners moved the company to north Denver.)

Some might attribute the decline in retail sales at Rocky Mountain Seed to the advent of regional upscale firms such as Seeds of Change and Seeds

Trust/High Altitude Gardens. Kenneth Vetting, the founder's grandson and recent owner-manager, offers another explanation: the growth of suburbia with smaller residential lots and thus less space for planting.

Gone are the fruit orchards, and traditional home vegetable gardens are seldom seen. And although sometimes hidden by the maze of unattractive highways and real estate developments, there has been a deliberate effort, through landscaping, to ameliorate the conditions for healthful living in the cities and suburbs of the Front Range. Leadership of this effort has come from local government, local voluntary groups, and the Cooperative Extension Service (CES) in partnership with the "green" industry. The "green" industry includes nurseries, greenhouses, sod farms, and retail garden centers as well as landscape architects, arborists, and landscape and other plant-care professionals.

Between 1950 and 1980, the population of the six metro Denver counties increased 263 percent, from 615,635 to 1,618,461. In view of these statistics, combined with the ever-present scarcity of water and the fact that nearly half of all domestic water consumption went for irrigating lawns, it is not surprising that the initiative for water-conserving landscaping came from the Denver Board of Water Commissioners. The immediate cause for action was an especially severe drought in the summer of 1977. After much conversation, Bill Miller, manager of Denver Water, as the board became popularly known, and Jim Grabow, president of the Associated Landscape Contractors of Colorado, established a joint committee on water conservation in early 1981. Its task was twofold: to create a demonstration garden, showing that by careful selection of plants using small amounts of water, one could have a successful and attractive yard; and to carry out a public education program that would stimulate and engage broad public support for "water-smart" landscaping. Notable members of the initial committee were John Wilder, conservation manager for Denver Water; Donald H. Godi, landscape architect; Larry Keesen, irrigation systems designer; Larry Watson, nurseryman; Gene Eyerly, horticulture consultant; and James R. Feucht, extension professor of horticulture at Colorado State University.

With financial contributions from local landscape contractors, the committee established a demonstration garden on one-third acre along the southwest side of Denver Water's administration building (between Eighth and Twelfth avenues, just east of I-25). Dedicated in spring 1982, the garden now surrounds the building and displays more than 200 varieties including trees, shrubs, ornamental perennials, and test plots of water-wise turf grasses.

To describe the landscaping approach used in the demonstration garden and to define that approach through an educational program, the commit-

Denver Water's demonstration xeriscape garden. Courtesy, Colorado WaterWise Council.

tee invented the word "xeriscape." Trademarked by Denver Water, xeriscape comes from "xeric," derived from the classical Greek word for arid conditions. The committee defined xeriscaping as "water conservation through creative landscaping" and adopted seven principles of xeriscaping that are still in use: plan and design from the start, create practical turf areas using drought-resistant grasses, select low-water-requiring plants, use soil amendments such as compost or manure, use mulches such as woodchips to reduce evaporation and keep the soil cool, irrigate efficiently, and properly maintain the landscape.[1]

Since the early 1980s, xeriscaping has spread rapidly. In Colorado, for example, demonstration gardens can now be found not only along the Front Range but also in several small towns on the eastern plains; and local nurseries are featuring more and more plants requiring no irrigation once established. Beyond the High Plains, xeriscaping has become especially popular in California and the Southwest. "Xeriscape" has found its way into the *Oxford English Dictionary* (fifth edition, 2002): "a style of landscape design suitable for arid regions, which aims to minimize the need for irrigation and other maintenance by the appropriate choice of plants and other features; a garden or landscape designed in this way." The word may be new, but the approach is at least as old as classical Greece, itself an arid region.

To be sure, xeriscaping does not mean digging up lawns and replacing them with gravel. At elevations below 6,000 feet, it does mean replacing humid-climate Kentucky blue grass (*Poa pratensis* L.) with semiarid-climate buffalo grass (*Buchloë dactyloides* (Nutt.) Engelm.) or the newer blends of drought-resistant varieties of fescues and bluegrass. The Rocky Mountain Seed Company was the first commercial seed house to sell buffalo grass, from which it still can be purchased in small or large quantities.

Colorado Cooperative Extension Service stands at the forefront of carrying the xeriscape message. Since the mid-1950s, Colorado CES has successfully adapted its programs and activities to serve the horticultural interests and requirements of city dwellers, suburbanites, and, most recently, exurbanites living beyond suburbia in the counties. By the late 1990s, exurbanites with their one- to seventy-acre "ranchettes," many with horse corrals, made up the fastest-growing segment of CES clientele.[2]

Positioning Colorado CES as a leader in urban horticulture was the major task of James R. Feucht, Colorado State University's representative to the 1981 xeriscape committee. A Denver native, Feucht had worked as a summer field assistant at the Cheyenne Horticultural Field Station during the 1950s, graduated from Colorado State University, and earned a Ph.D. in horticulture (arboriculture) at Michigan State University. He taught in the East before returning to Colorado State in 1966. Feucht was headquartered in Denver rather than on the Fort Collins campus until his retirement in 1994. He served as specialist to urban-county CES agents trained as horticulturists, not as agriculturists as was the case in the rural counties. He worked closely with the green industry, conveying campus research as well as training industry workers in landscape management. For the general public, he wrote a weekly column in the *Rocky Mountain News*.

During Feucht's tenure as extension professor, the population of metro Denver more than doubled. He provided this most recent wave of immigrants with information on what early settlers had learned through trial and error, namely, that "gardeners who are patient, know how to select plants that will do well, and manipulate the soil and microclimate will be amply rewarded." The undeniable fact, in his judgment, is that "Colorado grows the nation's best lawns, top-quality cut flowers and excellent vegetables." Newcomers needed to recognize that in Colorado's arid climate, "low humidity, fluctuating temperatures, heavy calcareous soils and drying winds often restrict plant growth more than low temperatures [do]." That answers a question often raised by immigrants from northern states such as Minnesota and Wisconsin, who wonder why trees that grow well back home do so poorly in Colorado.[3]

Feucht outlined the major obstacles to successful gardening along the Front Range and provided useful suggestions. To begin with, the region's notoriously clayey soils make for poor aeration, thus limiting the amount of oxygen needed for root growth. Simply adding water makes the soils even heavier, so Feucht recommended adding organic material such as peat moss (sphagnum) and manure to make them lighter. While a moderate amount of organic material improves soils, a large amount leads to the gradual accumulation of natural salts. Such salts can damage roots under conditions of rapid evaporation, when soils lose the porosity to allow salts to leach away. For all except the most extreme conditions of salinity, evidenced by the whitish, calcified alkali flats scattered on the High Plains, Feucht recommended the selection of plants tolerant of soils with an acidity/alkalinity measurement greater than pH 6.5. (Neutral is pH 7; less than pH 7 is acid, more than pH 7 is alkaline.)

Complicating the region's soil conditions is the presence of high amounts of iron, as evidenced by the dominant red color of the soil. While plants do require iron, the iron contained in the soils is in a form plants cannot absorb, which explains why so many non-native plants suffer from iron deficiency, or chlorosis, as exemplified by the yellowing of their leaves. The preferred solution, again, is to use only alkaline-tolerant plants.

Newcomers must also recognize that on the High Plains there are no "normal" years or "average" seasons. Heavy late-spring snows can break tree limbs, making plants susceptible to disease; subfreezing temperatures can destroy blossoms, preventing fruiting if not outright destroying plants. The choice is between fast-growing, brittle trees and slow-growing, limber trees and between plants generally native or acclimatized, which bloom late, and plants generally from milder climates, which bloom early. Similarly, for early-fall frosts, if we choose warm-climate plants such as the tomato, which cannot be put out until all danger of spring frost is past, it is problematic whether we can harvest before the first frost in the fall. Growing seasons along the Front Range vary from 120 days in Cheyenne to 144 days in Fort Collins and 162 days in Denver. In contrast, Laramie averages 75, Scottsbluff 131, and Hays 172 frost-free days.[4]

Consistent with the trend toward urban horticulture, Colorado became the third state in the nation to join the Master Gardener program, started in 1972 by David Gibby, CES agent in King and Pierce counties (Seattle), Washington. Introduced to metro Denver and organized by Feucht and urban horticulture extension agents, the program trained and certified volunteers to assist CES staff in providing reliable information to home gardeners. Besides answering questions called into extension offices, Master Gardeners taught horticulture to garden and civic clubs, helped with tree plantings sponsored by the schools, and assisted with community beautification and community gardening projects. In recent years, Metro Denver Master Gardeners have helped with "Earth Gardens," a variation of Helen Grenfell's plan for elementary students to learn about the natural sciences through direct participation. Roughly one-third of the inquiries received by Master Gardeners come from home owners seeking second opinions on recommendations made by commercial nurserymen and others in the green industry. In 2004, more than 1,700 Master Gardeners volunteered throughout the state of Colorado.[5]

Since 1997 Colorado CES and the Denver Botanic Gardens jointly have issued annual lists of plants best adapted to gardening and landscaping not only along the Front Range but throughout the Rocky Mountain Empire. Initiated by James E. Klett at Colorado State University and Panayoti Kalaidis at the Botanic Gardens, PlantSelect lists come in three categories: *recommended* plants, long grown in the region but, in the opinion of the cosponsors, yet to attain the popularity they deserve; *originals*, selected cultivars bred and tested by the cosponsors for superior performance; and *introductions*, varieties offered by commercial "cooperators." In 2005, retail nurseries participating in the PlantSelect program numbered seven on the Front Range and two beyond; wholesale nurseries numbered eleven in Colorado, one in Nebraska, and thirteen beyond the High Plains.[6]

In addition to using publications, seminars, radio, and television programs to convey horticultural information, Colorado CES sponsors a Web site that contains nearly 200 "Gardening Online Fact Sheets" that cover every aspect of horticulture—vegetables, fruits, ornamentals, trees and shrubs, insects and diseases, and irrigation. Not long ago, gardening was unrefined and uncomplicated, producing excellent edibles and attractive yards while providing exercise and enjoyment. It now seems that the farther away we get from our rural roots, the more rational and complex our gardening becomes.

Beyond providing horticultural information and technical assistance to individual residents, Colorado State University has extended the agricultural experiment station mission in support of the burgeoning green industry at many levels: through research the industry cannot afford to carry out on its own; training graduates in botany and horticulture for jobs in the industry; and operating specialized facilities such as the State Seed Laboratory, which provides testing for commercial applications and monitoring of the Colorado [Pure] Seed Act. Urban horticulture, in other words, has developed into a big business. In Colorado alone, the green industry in 2002 provided over 34,000 jobs and contributed over $2 billion in goods and services to the state's economy.[7]

Urban horticulture now focuses more and more on what historically we would have called ornamental horticulture and depends increasingly upon the manipulation of genetic material to improve the hardiness of plants. Today's genetic engineering accelerates the evolutionary process far more quickly than Niels Hansen could have possibly imagined. At the same time that horticulture as a profession has become more technical, our region as well as the rest of the nation has seen the flowering of contemporary planning professions. Whether through the 1,200-plus members of the Urban Land Institute, Colorado District, or through the more local "smart growth" citizens' initiatives, the emphasis is on development that as far as possible preserves the landscape while promoting sustainable communities.

In the area of urban horticulture, meanwhile, Fort Collins is unsurpassed. Judged among the very best small cities in America, Fort Collins is home to a population of about 140,000, to Colorado State University, and to five wholesale and retail PlantSelect growers. As a result of highly refined and comprehensive planning, the municipality has established and enforces codes to ensure well-landscaped residential and commercial areas. The municipal Forestry and Horticulture Division maintains more than 40,000 shade trees, over 600 acres of parks, 5,000 acres of natural areas, and 20 miles of off-street hiking and biking trails. In 1995 the city council approved a community horticulture program that included numerous gardens and other gardening activities. Two years later, voters approved the expenditure of $3 million for the development of "The Gardens on Spring Creek," an 18-acre horticultural site in southwest Fort Collins.

In 2005 The Gardens contained a greenhouse and classroom facility, with work under way for a "teaching" kitchen, an indoor garden or conservatory, and a children's garden. Thirty-five garden plots were available for minimal rental to individuals and families wishing to grow their own vegetables or ornamentals. Plans also call for the creation of ornamental "theme" plots, orchard groves and vegetable patches, and a formal garden. Altogether, The Gardens are designed to demonstrate how horticulture can improve urban living.[8]

The precursor of The Gardens was the community horticulture program organized as the Cheyenne Botanic Gardens, just forty-three miles north of Fort Collins but a significant 1,059 feet higher, with a full inch less annual precipitation (13.31 versus 14.47 inches) and more exposure to climatic extremes. Again, Cheyenne provides the ultimate test site for horticulture on the High Plains.

Community horticulture at Cheyenne began in 1977 with a solar-heated greenhouse on the eastern edge of the city. Supported by federal antipoverty funds, its purpose was both social and economic: to enable the elderly, troubled youth, and handicapped volunteers to grow plants year-round. In 1986 the program moved into a solar-heated indoor garden newly constructed at Lions City Park in north-central Cheyenne; together with the surrounding eight acres, the Cheyenne Botanic Gardens became a division of the City Parks and Recreation Department. The indoor garden is divided into three greenhouses, for community vegetable gardening, for starting ornamentals to be transplanted into municipal flowerbeds and parks, and for more conventional botanical displays. In keeping with its community service origins, over 90 percent of all work continues to be performed by volunteers.[9]

The same antipoverty agency that created the Cheyenne Botanic Gardens has sponsored a farmers market in downtown Cheyenne since 1980. Each Saturday from August through October, the market attracts thirty to forty vendors from northern Colorado, western Nebraska, and eastern Wyoming, as well as hundreds of area buyers. Farmers markets began in Torrington in 1999, sponsored by Wyoming Master Gardeners, and in Laramie in 2002. While reestablished more recently in Wyoming than in the other High Plains states, the success of all farmers markets suggests a public yearning for high-quality fruits, vegetables, and ornamentals not otherwise available and nostalgia for a simpler time of small, independent farms, when produce was grown with few or no artificial fertilizers, pesticides, and herbicides.

Cheyenne also continues its century-long community forestry tradition. With a population of about 55,000, the municipality maintains an Urban Forestry Division that in 2005 employed two full-time professional foresters, five arborists, and four seasonal workers. Besides caring for 17,000 municipal trees (cottonwood is still the most common species), Urban Forestry provides information to citizens and tree-care professionals and advises private developers and city crews on preferred tree-planting locations and species selections along new and existing rights-of-way. Beginning in 2005, Urban Forestry irrigated municipal trees with wastewater from the city's water treatment plant by means of a 5,000-gallon tanker trailer.[10]

In point of fact, Cheyenne is only one of more than fifty communities on the High Plains designated as a Tree City USA by the National Arbor Day Foundation of Nebraska City, Nebraska. These communities range in size from Brule (pop. 411 in 2002) in western Nebraska, Wiggins (838) in northeastern Colorado, Atwood (1,258) in northwestern Kansas, and Wheatland (3,548) in southeastern Wyoming to the major cities along Colorado's Front Range. To qualify as a Tree City USA, each community must maintain a volunteer tree board or municipal forestry department, pass a tree-care ordinance, invest at least two dollars annually per citizen in community trees, and celebrate Arbor Day.[11]

Designation as a Tree City USA does not ensure, however, that a municipality fully embraces an overall beautification mission. On the other hand, as the smaller towns especially on the High Plains seek to secure their very survival, community forestry becomes more than a matter of aesthetics. It is well documented that trees increase real estate values by as much as 10 percent and contribute to general economic viability.[12]

To the traveler across time, the most evident impact of tree planting is seen in Nebraska. When settlement began in the early nineteenth century, trees covered fewer than 1 million acres, or 2 percent of Nebraska's land surface. By the turn of the twenty-first century, trees covered more than 2 million acres, or 4 percent of the state. Such growth has been the result of deliberate suppression of wildfires, protection of riparian areas, reversion of marginal croplands to groves, and continual planting of trees. Today, virtually every Nebraska farmhouse is surrounded by windbreaks, generally featuring the red cedar (*Juniperus virginiana* L.), which in some places on the High Plains has actually changed the nature of rangelands.

Along Nebraska highways, windbreaks serve as "living snow fences" as well as elements of roadside beautification, most notably along Interstate 80 from North Platte to the Wyoming state line. Since the National Recovery

Act of 1934, at least 1 percent of federal funds apportioned to the states for highway construction needs to be spent for roadside beautification, which the act defines as preservation of scenic features, mitigation of soil erosion, seeding of grasses, and planting of trees and shrubs.

Through the Roadside Improvement Unit, Construction Division, the Nebraska Department of Roads hired its first landscape engineer in 1934 and built the first roadside park (along U.S. Highway 20 near Valentine) using federal funds. By the 1960s the Department of Roads was contracting with the University of Nebraska, Department of Horticulture, for assistance in identifying the best native species for roadside planting. Toward that end, horticulture majors from Lincoln were sent to collect in the eight westernmost counties, where conditions for growing were clearly the most difficult to accommodate.[13]

To further encourage tree planting and to help improve community appearances, the Nebraska Forest Service, a unit of the Institute of Agriculture and Natural Resources at the University of Nebraska, established the nation's first and only statewide arboretum during the 1980s. Functioning as a "network of arboreta, parks, historic properties and other public landscapes," the arboretum is supported financially by the university, a charitable nonprofit membership organization, and the sponsors of its numerous community affiliates. The affiliates are classified into three categories: "accredited arboreta," where plants are selected and cultivated for specified educational purposes; "historic landmarks," planned landscapes with historic or cultural significance; and "landscape steward" sites that demonstrate "an ongoing commitment to excellence in design, plant diversity and maintenance."[14] Affiliates west of the 100th meridian include the arboreta at Gering, Scottsbluff, and North Platte; the landmarks of the Nebraska National Forest near Halsey and Harmon Park at Kearney; and landscapes at Alliance, Cambridge, and Kearney.

Although not a formal part of Nebraska CES, the statewide arboretum contributes to the overall land-grant college mission by applying useful knowledge from campus to the communities. Its Community Landscape Program alone has provided assistance to over 150 municipalities in improving the appearance of their parks, school yards, and other public places.

In conjunction with the Nebraska Nursery and Landscape Association, the statewide arboretum sponsors GreatPlants, a program to select and promote the distribution of ornamentals suitable to the various regions of Nebraska. As the 2006 "tree of the year," GreatPlants recommended the Chinkapin, or yellow chestnut oak (*Quercus muehlenbergia* Engelm.), a native of southeastern Nebraska but recommended for the High Plains because of its relative alkaline and drought tolerance; as "perennial of the year," the pasque flower (*Pulsatilla patens* P. Mill. [*Anemone patens* L.]), a High Plains native, among the earliest spring flowers, and South Dakota's state flower; and as "grass of the year," autumn red miscanthus (*Miscanthus* "Purpurascens"), originally from East Asia, valued for its red-orange fall leaf color and its cold-hardiness. The statewide arboretum works together with specialized nurseries in a continuing worldwide search for plants adaptable to the Great Plains.[15]

In Kansas, too, the land-grant college supports the green industry in promoting all forms of horticulture. For example, Kansas CES and the Kansas Nursery and Landscape Association jointly sponsor "Pride of Kansas Plants." Each year, Kansas CES publishes a list of recommended trees, shrubs, annuals, and perennials. In 2005, the list featured bur oak (*Quercus macrocarpa* Michx.), native to central Kansas but found adaptable to the High Plains, and dwarf Oregon grape (*Mahonia Aquifolium Compactum*), a cross between the native Oregon grape (*Berberis repens* Lindl.) and an Asiatic barberry, which grows as a mound-shaped shrub up to a height of three feet. In addition, Kansas CES sponsors "Prairie Star Flowers," a selection of annuals tested at four Kansas State University research sites, among them Hays and Colby. Although Kansas CES publishes a wide variety of horticultural "Tip Sheets" and sponsors Master Gardener and other horticultural programs, its efforts appear almost entirely aimed at the more populated eastern section of the state. So, too, with the numerous statewide associations—landscape architects, vegetable and fruit growers, arborists, florists, sod growers, and others—which emerged from, and by 1980 had replaced, the State Horticultural Society.[16]

If there is one particular area on the High Plains where history has recorded definitively the inability of the land to sustain traditional agriculture, much less traditional horticulture, it is western Kansas. As noted earlier, homesteaders could survive with a combination of wells, windmills, and small reservoirs, but they could not rely on natural precipitation to produce the crops that brought the income that enabled them to thrive. As large-scale irrigation around Garden City, using primarily Arkansas River water, proved unreliable, private investors turned to a new, seemingly inexhaustible source of "underflow," or groundwater. The prospect of drawing water from a vast underground river gained credence as the result of a United States Geological Survey report in 1891 and pretty much remained an accepted truth until the 1960s, when groundwater levels began showing marked declines.[17]

Taken as a whole, the groundwater in question is known as the Ogallala aquifer, the water-bearing stratum of permeable rock created millennia ago by underground runoff from the Rocky Mountains. While the Ogallala aquifer underlies most of the High Plains, the most dramatic declines in water levels have occurred at the greatest distance from the mountains: in southeastern Colorado and western Kansas, roughly from just beyond the Arkansas River valley north to Interstate 70. While some still argue publicly about the rate of decline, everyone dependent on the aquifer knows that current levels of use are insupportable. The implications for agriculture, horticulture, and communities are immeasurable.

Since 1945, Kansas has been a "prior appropriation" state, meaning that all water, including underground water, is a public resource allocated by the state engineer. To address the issue of aquifer decline, and thereby to assist the state engineer, the 1972 Kansas Legislature authorized the establishment of locally organized groundwater management districts. Kansas Groundwater Management District #4 covers all of three counties and portions of seven more on the High Plains. In 1990 the district board decided to adopt a program leading to zero depletion by 2000. While the state engineer approved, district voters expressed substantial opposition at public hearings. So the district withdrew its zero-depletion plan in favor of an incentive program. Through the Northwest Kansas Groundwater Conservation Foundation, a charitable nonprofit corporation, the district can provide financial incentives from a trust fund to encourage water users either to reduce their irrigated acreage or otherwise to reduce evaporation and transpiration. The foundation debated whether to purchase or lease water rights over the most threatened areas, consistent with the state's revised goal, adopted in 2004, for "slowing the decline rate and extending the life of the aquifer."[18]

Similarly, in northeastern Colorado, farmers and communities are seeking ways, so far voluntary, to reduce the amount of water extracted from the Ogallala aquifer. To comply with the final settlement of the Republican River Compact (December 2002), Governor Bill Owens signed an act of the 2004 Colorado Legislature to establish the Republican River Water Conservation District, which, in turn, has chosen the Conservation Reserve Enhancement Program through the United States Department of Agriculture as the primary incentive to reduce the amount of irrigated land. While the federal program supports taking irrigated land out of temporary production, groups such as the Nature Conservancy are seeking to pay farmers to permanently close their wells.[19]

The decline in groundwater level has added to the problem of conserving the very land that has proven the most difficult to protect from desertification. We made reference earlier to the experimental work, most notably on buffalo grass, at the Hays substation following the Dust Bowl, stating that buffalo grass continues to be used for revegetation of grasslands and, to an increasing extent, in residential and commercial landscaping. In fact, the principal of planting a perennial that restores, or at least does not deplete, the soil suggests a newer, more supportable attitude toward horticulture generally. To some extent, Niels Hansen revealed that attitude when he argued that the horticulturist's role is to "work with Nature[,] not against" and that plant breeding is simply a technique to expedite the work of nature: "Acclimation is Nature's work, extending through thousands of years. Acclimatization—Man's work—is useful as in shortening the season."[20]

The disappearance of an estimated one-fourth to one-third of Kansas topsoil within the past 150 years strongly suggests that we have been working against, not with, nature. To reverse that trend, Wes Jackson, a Kansas native, founded the Land Institute near Salina in 1974. His idea is to develop "Natural Systems Agriculture" through which perennials that produce high yields of seeds would replace wheat and other grass annuals. Jackson, trained as a geneticist, and his cadre of scientists carry out the core of their work on the institute's 580 acres, which consist of farmland on both banks of the Smoky Hill River, upland farmland, nursery plots, and 200 hilly acres of native and restored prairie. In 2004 the institute worked on crossing a wild wheatgrass with cultivated wheat to create an economically viable perennial crop requiring less labor, less water, and no soil amendments. While Salina is two degrees longitude east of the High Plains and Jackson's interest is agricultural, the potential application of the institute's work for both horticulture and community on the High Plains is quite radical, if not entirely practical. Some view Jackson as a hopeless idealist because of his conviction that someday the family farm could thrive on a "polyculture of herbaceous perennials" rather than on the current monoculture of annuals. On the other hand, major migrations out of western Kansas illustrate the land's inability to sustain agriculture, much less communities, despite dramatic advances in irrigation technology. John Wesley Powell had it right in 1878 when he concluded that, except for a few valleys, the vast portion of the arid West, including the High Plains, was not irrigable.[21]

In recognizing the limitations of the land and the settlement history of the High Plains, Wes Jackson and his cadre are part of an imaginative body of researchers and practitioners forging new steps to reach the interconnected goals of agricultural and small-town sustainability. Given the economic and non-economic benefits of beautification, one may reasonably conclude that horticulture can further the revitalization of rural communities just as it has contributed, and continues to contribute, to the general well-being of cities.[22]

Notes

1. Colorado Waterwise Council, "Xeriscape Colorado," at http://www.xeriscape.org; accessed May 2005. James H. Feucht prepared five Colorado State University Cooperative Extension Gardening Online Fact Sheets on xeriscaping, revised August 2004: "Creative Landscaping," at http://www.ext.colostate.edu/pubs/garden/07228.html; "Trees and Shrubs," at http://www.ext.colostate.edu/pubs/garden/07229.html; "Ground Cover Plants," at http://www.ext.colostate.edu/pubs/garden/07230.html; "Garden Flowers," at http://www.ext.colostate.edu/pubs/garden/07231.html; and "Retrofit Your Yard," at http://www.ext.colostate.edu/pubs/garden/07234.html; all accessed October 2005.

2. Colorado State University Cooperative Extension Service Annual Report 68 (1955), and 5 (1998): 17, Colorado State University Library—Archives.

3. James R. Feucht, "Colorado Gardening: Challenge to Newcomers," Colorado State University Cooperative Extension Gardening Online Fact Sheet, 2001, 1, at http://www.ext.colostate.edu/pubs/garden/07220; accessed May 2005.

4. Ibid., 1–3.

5. Colorado State University Cooperative Extension Service Annual Report (1997): 10, Colorado State University Library—Archives.

6. http://www.plantselect.org; accessed October 2005.

7. http://www.greenco.org; accessed October 2005.

8. http://www.fcgov.com/horticulture; accessed May 2005.

9. http://www.botanic.org; accessed October 2005.

10. http://www.cheyennetrees.com; accessed May 2005; Paula J. Peper, E. Gregory McPherson, James R. Simpson, Scott E. Maco, and Qingfu Xiao, "City of Cheyenne, Wyoming Municipal Tree Resource Analysis," prepared by the Center for Urban Forest Research, USDA Forest Service, Pacific Southwest Research Station, Davis, California (2004), 62 pp.

11. http://www.arborday.org; accessed October 2005.

12. N. Robin Morgan and Kenneth J. Johnson, "An Introductory Guide to Urban and Community Forestry Programs," USDA Forestry Report R8-FR16 (October 1993): 1.

13. George E. Koster, *A Story of Highway Development in Nebraska* (Lincoln: Nebraska Department of Roads, 1997), 38–39; Sotero S. Salac, "Collection, Propagation, Culture, Evaluation, and Maintenance of Plant Materials for Highway Improvement," research report prepared by the Department of Horticulture, University of Nebraska (Lincoln: Nebraska Department of Roads, 1977), 185 pp.

14. http://arboretum.unl.edu/affiliate/index.html; accessed October 2005.

15. http://arboretum.unl.edu; accessed October 2005.

16. Alan Stevens, "Prairie Star Flowers" [flyer], Department of Horticulture, Forestry and Recreation Resources, Kansas State University, Manhattan; E. P. Lana, ed., *A History of Horticulture of the Great Plains Region, 1960 through 1981* (St. Paul: North Central Region, American Society for Horticultural Science, 1984), 25–27.

17. John Opie, *Ogallala: Water for a Dry Land* (1993; 2nd ed., Lincoln: University of Nebraska Press, 2000), 60–74.

18. Ibid., 210, 216; James R. Dickenson, *Home on the Range: A Century on the High Plains* (Lawrence: University of Kansas Press, 1995), 248; quote from Wayne Bossert, general manager, Kansas Ground Management District 4, e-mail message to author, May 2, 2005; Robert G. Dunbar, *Forging New Rights in Western Water* (Lincoln: University of Nebraska Press, 1983), 189.

19. Amanda Paulson, "To Reduce Water Use, Farmers in Colorado Tax Themselves," *Christian Science Monitor*, March 19, 2007, at http://www.csmonitor.com/2007/0319/p02s01-ussc.htm; accessed March 2007.

20. Niels E. Hansen, "Fifty Years Work as Agricultural Explorer and Plant Breeder," *Iowa State Horticultural Society Transactions* 79 (1944): 33.

21. Wes Jackson, *New Roots for Agriculture*, 2nd ed. (Lincoln: University of Nebraska Press, 1985), 93, 99–100; http://www.landinstitute.org; accessed May 2005.

22. Foremost among contemporary proponents of the economic benefits of beautification is Edward T. McMahon, senior fellow at the Urban Land Institute and formerly with the Conservation Fund.

Postscript

The development of horticulture on the High Plains fits into Jefferson's theme of the "march of civilization," making "barren" land more amenable to human settlement; it also suggests some lessons for those more rural communities that have yet to embrace the economics of beautification.

At the beginning of our story, when soldiers at Fort Laramie were growing vegetables for their physical sustenance, Andrew Jackson Downing, dean of American landscape architecture, was discoursing to his landed eastern readers on the connection between citizenship and horticulture. He explained that the more attractive the home surroundings, the more enjoyable life becomes and the more attached we get to our locale: "The love of country is inseparably connected with the love of home." Landscape gardening, he continued, is good in another way: "There is no employment or recreation that affords the mind greater or more permanent satisfaction, than that of cultivating the earth and adorning our own property. 'God Almighty first planted a garden; and, indeed, it is the purest of human pleasures,' says Lord Bacon. And as the first man was shut out from the garden . . . the desire to return to it seems to be implanted by nature, more or less strongly, in every heart."[1]

Whether the missionaries for horticulture on the High Plains had read Downing's work is unclear. His moralistic message, however, would have resonated with teachers such as Charles Bessey of Nebraska and Aven Nelson of Wyoming, publicists such as Lucius Wilcox of *Field and Farm*, and preacher-horticulturists such as Charles S. Harrison of Nebraska. More than anyone, Harrison carried that message to every state on the High Plains, aiming his exhortations not so much at town dwellers as at homesteaders. "Many a man," he wrote, "has made his home yard a cattle-pen and a pig-sty." And "many a noble woman with love for the beautiful has died under the slow martyrdom of ugliness." How much better, then, to have a home surrounded by "supreme loveliness: a clump of evergreens here and of flowering shrubs there, graceful walks and well-kept beds of flowers."[2]

Referring to the horticulturist as "the high priest of Nature" and "co-worker with God," Harrison exhorted his fellow amateurs to "lift [horticulture] above the plans of hogs and corn and unfold its grand mission, that it may make the world healthier, better, and elevate it." God is the Creator; we horticulturists ask for His blessing and His help, but we know that on earth His work must truly be our own.[3]

"What we lost in Eden of yore, we as gardeners seek to restore." Some may have taken literally this motto of the South Dakota Horticultural Society; others such as Harrison would have used it metaphorically. Consistent with the latter interpretation, recent empirical research suggests that there is indeed a connection between living well and attractive surroundings. Edward O. Wilson, a Harvard entomoloist, has used the word "biophilia" to describe our instinct for and empathy with the natural world. Once humans manage to rise above the level of mere subsistence, he has argued, they expend great effort to make their surroundings more livable according to "aesthetic criteria." Given a choice, for example, humans will invariably select home sites on "open tree-studded land on prominences overlooking water."[4]

We may not think about the green industry as "responding to a deep genetic memory of mankind's optimal environment," but that is the reality.[5] With an apparently innate desire to live in attractive surroundings, at least since ancient times, we have enjoyed cultivating our gardens as a way to unwind and recreate our energies. Additionally, therapeutic horticulture is now big business at botanic gardens, senior centers, hospitals, and even prisons; and courses on the subject are offered at each of the land-grant universities serving the High Plains. The post surgeon at Fort Laramie may have known nothing about such empirical research, but he did recognize the value of a flower garden for his patients.

Whether the family that gardens together stays together can be debated, but there is no doubt that an attractive home tends to increase its residents' attachment to neighborhood and community. Given the human instinct for the aesthetic and the limitations of land and water in the High Plains region, it is just possible that by making very modest investments in community beautification, as has occurred in Fort Collins, we can help attract reasonable numbers of highly trained professionals and well-paid service workers to revitalizing small towns on the High Plains.

Proposals to save our small towns, of course, are not new. In Topeka in September 1931, Franklin D. Roosevelt outlined a federal plan meant especially for communities on the Great Plains west of the 98th meridian: move farmers off submarginal land and provide financial support for their resettlement on profitable land; restore the Great Plains to their native grasses; decentralize industry and make it serve as the foundation for a new, nonagricultural economy.[6]

While Roosevelt's plan never materialized, the idea of a federal bailout persisted. At its most extreme, in 1987, Deborah Popper, a graduate student at Rutgers University in New Jersey, put forward a scheme in which the federal government would buy up submarginal lands, convert them to a vast park-like commons, then fund a massive economic redevelopment under the auspices of a semiautonomous agency similar to the Tennessee Valley Authority. Lost in the ensuing local uproar over the suggestion of converting private to public property was the fact that decades of federal policy had not brought the most rural part of the High Plains closer to the establishment of sustainable communities.[7]

The suggestion that the High Plains be turned into a vast short-grass commons is not as far-fetched as it may seem. Paul Johnsgard of the University of Nebraska has noted that, to a considerable degree, the Sandhills of Nebraska "already operate as a natural prairie commons. The region's native flora and fauna are still essentially intact, its original ecology is still largely apparent, and because of meager population and scattered landownership patterns it has few interfering roads and fences."[8]

Nor is the notion of a post-agrarian economy on the High Plains entirely improbable, considering the limitations of the land. Take the example of Imperial, Nebraska—population 1,982 in 2000, elevation 3,271 feet, located on U.S. 6 about fifty miles south of Ogallala (along I-80), surrounded by stock farms and thus with some of the best pheasant hunting on the High Plains.

Town Park, Bird City, Kansas, 2007. Courtesy, Catherine Domsch.

A Tree City USA since 1984, Imperial provides incentives to residents who maintain their properties, including waving dump fees for inoperable vehicles—to some people, the universal bane of community beautification and, to others, the ultimate measure of personal possession and freedom from community constraint. In the early 1990s, residents young and old raised the money and planted an arboretum around the new high school. During recent high school reunions, returning alumni have expressed universal pleasure and amazement at how attractive the community has become. At one time, alumni who decided to return to Imperial did so upon retirement; however, within the past few years, younger alumni have returned with their families, thanks in part to the creation of jobs in a new, home-grown telecommunications company and in part to the visible results of small, inexpensive community-building projects. To provide current and former residents with an opportunity to voluntarily invest in community betterment and to ensure that needs yet unknown can be met in the future, a small group of residents established the Imperial Community Foundation, an affiliate of the Nebraska Community Foundation.[9]

Not far away, at Bird City, Kansas, one finds another example of folks determined not to allow their community to wither away. Bird City, population 482 in 2000, elevation 3,460 feet, is just off U.S. Highway 36, about forty miles northeast of Goodland (along I-70) and situated on a slight rise surrounded by vast wheat fields. In 2002, the town's former banker left a $5 million legacy, with instructions that earnings be spent for community development; by 2004, that legacy had become part of a local community foundation. Its earliest grants were for maintenance of town parks; since then, its board has decided to invest in building an attractively landscaped community gateway as well as in demolishing dilapidated buildings on the main street. This, too, is part of a new, inexpensive, and comprehensive effort to make small towns ready for business.[10]

In a sweeping and farsighted essay, Michael Lind of the New America Foundation has suggested that, by reducing direct and indirect subsidies to uneconomical farms and ranches, the federal government could pay for a high-technology infrastructure that, in turn, would support a combination of new or expanded service and manufacturing industries.[11] While there is little likelihood that Congress will allow the federal government to shift its subsidies any time soon, small towns such as Imperial and Bird City are taking the initiative to attract a new wave of high-technology pioneers. As that happens, small towns on the High Plains will become part of an archipelago of livable communities in a sea of short-grass.

Notes

1. Andrew Jackson Downing, *The Theory and Practice of Landscape Gardening Adapted to North America with a View to the Improvement of Country Residences* (1850; repr. Washington, D.C.: Dumbarton Oaks, 1991), ix.

2. C. S. Harrison, "The Ethics of Horticulture," Nebraska State Horticultural Society Annual Report (1896): 215.

3. Ibid., 209.

4. Edward O. Wilson, *Biophilia* (Cambridge: Harvard University Press, 1984), 108–111. See Michael Waldholz, "Flower Power: How Gardens Improve Your Mental Health," *Wall Street Journal*, August 26, 2003 (section D), for a summary of research by Wilson and by Roger Ulrich, professor of architecture and landscape architecture and director of the Center for Health Systems and Design at Texas A&M.

5. Wilson, *Biophilia*, 111.

6. Frank Friedel, *Franklin D. Roosevelt: Launching the New Deal* (Boston: Little, Brown, 1973), 4: 79.

7. Deborah E. Popper and Frank Popper, "The Great Plains: From Dust to Dust," *Planning Magazine* 53, no. 12 (December 1987): 12–18.

8. Paul A. Johnsgard, *This Fragile Land: A Natural History of the Nebraska Sand Hills* (Lincoln: University of Nebraska Press, 1995), 54.

9. Lori Pankonin, *Imperial Republican*, telephone conversation with author, October 28, 2005.

10. Catherine Domsch, Bird City Community Foundation, several conversations with author, 2005.

11. Michael Lind, "The New Continental Divide," *Atlantic Monthly* (January–February 2003): 86–88.

Bibliography

Abbot, Carl, Stephen J. Leonard, and David McComb. *Colorado: A History of the Centennial State*. Rev. ed. Boulder: Colorado Associated University Press, 1982.

Ahearn, M. F. "The Home Vegetable Garden." *Kansas Agricultural Experiment Station Circular* 64 (February 1918), 8 pp.

Alderman, William H. *Development of Horticulture on the Northern Great Plains*. St. Paul: Great Plains Region, American Society for Horticultural Science, 1962.

Allen, Martin. "Horticulture on the Plains." Kansas State Horticultural Society Annual Report 8 (1878): 187–189.

Annis, Frank J. "Concerning the Duties of the Secretary of the State Board of Agriculture, and the Distribution of College Seeds and Plants." *Colorado Agricultural Experiment Station Bulletin* 3 (December 1887): 1–4.

Aughey, Samuel. *Sketches of the Physical Geography and Geology of Nebraska*. Omaha: Daily Republican Book and Job Office, 1880.

Babb, Myron Francis. "Residual Effects of Forcing and Hardening on the Morphology and Physiology of Vegetable Plants." *USDA Technical Bulletin* 760. Washington, D.C. (December 1941), 35 pp.

Babb, Myron Francis, and James E. Kraus. "Home Vegetable Gardening in the Central and High Plains and Mountain Valleys." *USDA Farm Bulletin* 2000. Washington, D.C. (1949), 98 pp.

———. "Orach, Its Culture and Use as a Greens Crop in the Great Plains Region." *USDA Circular 526*. Washington, D.C. (September 1939), 23 pp.

———. "Results of Tomato Variety Tests in the Great Plains Region." *USDA Circular 533*. Washington, D.C. (1939), 12 pp.

Babb, Myron Francis, and W. L. Quayle. "Vegetable Culture and Varieties for Wyoming." *Wyoming Agricultural Experiment Station Bulletin* 250 (April 1942), 40 pp.

Bailey, Joseph Cannon. *Seaman A. Knapp: Schoolmaster of American Agriculture*. New York: Columbia University Press, 1945.

Bailey, Liberty Hyde. *The Country-Life Movement in the United States*. New York: Macmillan, 1920 [1911].

———. *Sketch of the Evolution of Our Native Fruits*. New York: Macmillan, 1911.

———, ed. *Report of the Commission on Country Life*. New York: Sturgis and Walton, 1911.

Baker, Gladys L., Wayne D. Rasmussen, Vivian Wiser, and Jane M. Porter. *Century of Service: The First 100 Years of the United States Department of Agriculture*. Washington, D.C.: USDA, 1963.

Ball, Jeff. "The Versatile Osage-Orange." *American Forests* 106, no. 3 (Fall 2000): 60–62.

Benson, Maxine, ed. *From Pittsburgh to the Rocky Mountains: Major Stephen Long's Expedition, 1819–1820*. Golden, Colo.: Fulcrum, 1988.

Bessey, Charles E. *Botany for High Schools and Colleges*. 6th ed. New York: Henry Holt, 1889 [1880].

———. *Elementary Botanical Exercises for Public Schools and Private Study*. Lincoln: J. H. Miller, 1894.

———. "The Forests and Trees of Nebraska." Nebraska State Board of Agriculture Annual Report (1899): 79–102.

———. "High School Botany." *Science* 7 (1898): 266–267.

———. "Industrial Education." Nebraska State Board of Agriculture Annual Report (1885): 77–83.

———. "A Meeting-Place for Two Floras." *Bulletin of the Torrey Botanical Club* 14 (1887): 189–191.

———. "Natural Horticultural Regions of Nebraska." Nebraska State Horticultural Society Annual Report (1887–1888): 63–75.

———. "Notes on the Botany of the Strawberry." Nebraska State Horticultural Society Annual Report (1896): 237–240.

———. Papers, 1865–1915. Microfilm edition. University of Nebraska Archives, Lincoln.

———. "The Plan of Work in the Department of Horticulture in the Industrial College of the University of Nebraska." Nebraska State Horticultural Society Annual Report (1886): 108–112.

———. "A Preliminary Report upon the Native Trees and Shrubs of Nebraska." *Nebraska Agricultural Experiment Station Bulletin* (1892): 171–198.

———. "Progress of the Botanical Survey of Nebraska." *American Naturalist* 29 (1895): 580–582.

———. "The Reforesting of the Sand Hills." Nebraska State Board of Agriculture Annual Report (1893): 94–97.

———. "Report of the Botanist upon the Grasses and Forage Plants of Nebraska." Nebraska State Board of Agriculture Annual Report (1888): 131–142.

———. "The Russian-Thistle in Nebraska." *Nebraska Agricultural Experiment Station Bulletin* 31 (1893): 67–77.

Bessey, Charles E., and Albert F. Woods. "The Botany of the Apple Tree." Nebraska State Horticultural Society Annual Report (1894): 7–36.

Binkley, Almund M. "The Home Vegetable Garden." *Colorado Agricultural Experiment Station Bulletin* 354 (February 1930), 50 pp.

Blount, A. E. "Report of Experiments with Grains, Grasses and Vegetables on the College Farm, 1887." *Colorado Agricultural Experiment Station Bulletin* 2 (1887), 16 pp.

Bowman, Albert E. "History of the Agricultural Extension Work in Wyoming" (typescript). Laramie: Wyoming Agricultural Extension Service, 1964, 76 pp.

Boyd, David. *A History: Greeley and the Union Colony*. Greeley: Greeley Tribune Press, 1890.

Boyd, George W., and Burton W. Marston. *The Wyoming Agricultural Extension Service and the People Who Made It, 1919–1964*. Laramie: University of Wyoming Publications, 1965.

Brandon, J. F., and Alvin Keezer. "Soil Blowing and Its Control in Colorado." *Colorado Agricultural Experiment Station Bulletin* 419 (January 1936), 20 pp.

Brengle, K. G. *Principles and Practices of Dryland Farming*. Boulder: Colorado Associated University Press, 1982.

Briggs, Lyman J., and H. L. Schantz. "The Water Requirements of Plants. I. Investigations in the Great Plains in 1910 and 1911." *USDA Bureau of Plant Industry Bulletin* 284. Washington, D.C. (1913), 49 pp.

Bryant, James Edwin. *What I Saw in California, Containing the Complete Original Narrative and Appendix from the 1849 Appleton Edition in True Facsimile*. Palo Alto: Lewis Osborne, 1967.

Bryant, O. W. "Progress Report on Experiments in Supplemental Irrigation with Small Water Supplies at Cheyenne and Newcastle, Wyoming, 1905–1909." *USDA Office of Experiment Stations Circular* 92. Washington, D.C. (January 1910), 51 pp.

Budd, Joseph L., and Niels E. Hansen. *American Horticultural Manual*. 2 vols. New York: John Wiley and Sons, 1902–1903.

Byers, William N. "Shade Trees." Colorado Board of Horticulture Annual Report 10 (1898): 86–90.

Call, Leland E., and Louis C. Aicher. "A History of the Fort Hays (Kansas) Branch Experiment Station 1901–1962." *Kansas Agricultural Experiment Station Bulletin* 453 (May 1963), 110 pp.

Campbell, Hardy W. *Campbell's 1907 Soil Culture Manual.* Lincoln: Campbell Soil Culture Co., 1909.

Candolle, Alphonse de. *Origin of Cultivated Plants.* 2nd ed. New York: Hafner, 1967 [1886].

Carey, Joseph M. "The Future of Horticulture in the State of Wyoming." *Wyoming State Board of Horticulture Special Bulletin* 1 (1907): 19–26.

Cather, Willa. *My Ántonia.* 1918. Boston: Houghton Mifflin, 1977.

———. *O Pioneers!* 1913. Boston: Houghton Mifflin, 1988.

[Cheyenne Horticultural Field Station.] Annual Report, FY1929, 1930, 1932, 1933, 1934, 1935, 1936, followed by Research Line Project Annual Report of Progress, FY1937–1945 (carbon copies). Cheyenne: USDA High Plains Grassland Research Center.

Chilcott, E. C. "Dry-Land Farming in the Great Plains Area." *USDA Yearbook* (1907): 451–468.

Condra, George E. "Tree Planting and Landscape Beautification in Nebraska." *Conservation and Survey Division Bulletin* 2, Lincoln (March 16, 1929), 29 pp.

Crandall, Charles S. "Notes on Plum Culture." *Colorado Agricultural Experiment Station Bulletin* 50 (1898), 48 pp.

———. "The Russian Thistle." *Colorado Agricultural Experiment Station Bulletin* 28 (1894), 12 pp.

———. "Strawberries." *Colorado Agricultural Experiment Station Bulletin* 53 (1900), 27 pp.

Creigh, Dorothy W. *Nebraska, a Bicentennial History.* New York: W. W. Norton, 1977.

Cunningham, Isabel S. *Frank N. Meyer, Plant Hunter in Asia.* Ames: Iowa State University Press, 1984.

Cunningham, J. C. "Protecting Trees from Rabbits." *Kansas Agricultural Experiment Station Circular* 17 (February 1911), 4 pp.

Curley, Edwin A. *Nebraska 1875: Its Advantages, Resources and Drawbacks.* Lincoln: University of Nebraska Press, 2006.

Dick, Everett N. *Conquering the Great American Desert: Nebraska.* Lincoln: Nebraska State Historical Society, 1975.

Dickenson, James R. *Home on the Range: A Century on the High Plains.* Lawrence: University of Kansas Press, 1995.

Dorsett, Lyle W., and Michael McCarthy, *The Queen City: A History of Denver.* 4th ed. Boulder: Pruett, 1986.

Downing, Andrew Jackson. *The Theory and Practice of Landscape Gardening Adapted to North America with a View to the Improvement of Country Residences.* 4th ed. Washington, D.C.: Dumbarton Oaks, 1991 [1850].

Dunbar, Robert G. *Forging New Rights in Western Waters*. Lincoln: University of Nebraska Press, 1983.

———. "History of [Colorado] Agriculture." In *Colorado and Its People*, ed. Leroy Hafen, 2: 121–157. New York: Lewis Historical Publishing, 1948.

Edmondson, William O. "Trees for Protection and Profit." *Wyoming Agricultural Extension Service Circular* 116 (April 1951), 39 pp.

———. "Trees Improve Your Farm." *Wyoming Agricultural Extension Service Circular* 27 (February 1947), 28 pp.

Emmons, David M. "Richard Smith Elliott, Kansas Promoter." *Kansas Historical Quarterly* 36, no. 4 (Winter 1970): 390–401.

Ewan, Joseph A. *Biographical Dictionary of Rocky Mountain Naturalists*. Utrecht: Bohn, Scheltema and Holkema, 1981.

Fairchild, David. *The World Was My Garden: Travels of a Plant Explorer*. New York: Charles Scribner's Sons, 1938.

Feucht, James R. "Colorado Gardening: Challenge to Newcomers." Fort Collins: Colorado State University Cooperative Extension Service Gardening Online Fact Sheet, 2001.

———. "Xeriscaping." Fort Collins: Colorado State University Cooperative Extension Service, *Service in Action Bulletin* 7, n.d., 228–232.

Filinger, George A. "The Kansas State Horticultural Society, 100 Years of Progress, 1867–1968" (mimeograph). Manhattan: Kansas State Horticultural Society, [1968].

Friggens, Paul. "Plant Pioneer of the Great Plains" [1949] (typescript). Box 2, folder 5, Hansen Papers. South Dakota State University Archives, Brookings, 5 pp.

Gleason, Henry A. *The New Britton and Brown Illustrated Flora of the Northeastern United States and Adjacent Canada*. New York: New York Botanical Garden, 1952.

Gordon, John H. "Experiments in Supplemental Irrigation with Small Water Supplies at Cheyenne, Wyoming in 1909." *USDA Office of Experiment Stations Circular* 95 (Washington, D.C.) (April 1910), 11 pp.

Greb, Bentley W. "Significant Research Findings and Observations from the U.S. Central Great Plains Research Station and Colorado State University Experiment Station Cooperating, Akron, Colorado, Historical Summary, 1900–1981" (mimeograph). Akron: U.S. Central Great Plains Research Station, 1981, 24 pp.

Greb, Bentley W., and D. W. Robertson. "Fifty Years of Agricultural Progress, 1907–1957, USDA, Colorado State University, and Local Cooperating Groups, June 28, 1957." Akron, Colo.: USDA Central Great Plains Field Station, 1957.

Gregg, Josiah. *Commerce on the Prairies*, ed. Max L. Moorhead. Norman: University of Oklahoma Press, 1954 [1844].

Grenfell, Helen L. "The Public Schools and Horticulture." Colorado Board of Horticulture Annual Report 13 (1901): 177–186.

Hafen, Leroy R., ed., *Colorado and Its People*. New York: Lewis Historical Publisher, 1948, 2 vols.

Hall, William K. "The Investigation Now Being Made in Nebraska by the U.S. Bureau of Forestry." *Nebraska State Horticultural Society Annual Report* (1902): 149–159.

Haney, J. G. "Experiments at Fort Hays Branch Station, 1902–04." *Kansas State Agricultural Experiment Station Bulletin* 128 (December 1904), 77 pp.

Hansen, Niels E. "Breeding Hardy Fruits." *South Dakota Horticultural Society Annual Report* 4 (1907), 32 pp.

———. "Breeding Hardy Strawberries." *South Dakota Agricultural Experiment Station Bulletin* 103 (June 1907): [216]–265.

———. "Experiments in Plant Heredity." *South Dakota Agricultural Experiment Station Bulletin* 237 (April 1929), 24 pp.

———. "Fifty Years Work as Agricultural Explorer and Plant Breeder." *Iowa State Horticultural Society Transactions* 79 (1944): 28–49.

———. Papers, 1888–1953. South Dakota State University Archives, Brookings.

———. "Plant Introductions." *South Dakota Agricultural Experiment Station Bulletin* 224 (May 1927), 64 pp.

———. "Plums in South Dakota." *South Dakota Agricultural Experiment Station Bulletin* 93 (May 1905), 88 pp.

———. "Questions and Answers on Fruit Culture." *South Dakota Agricultural Experiment Station Circular* 35 (June 1941), 30 pp.

———. "A Study of Northwestern Apples." *South Dakota Agricultural Experiment Station Bulletin* 76 (June 1902), 142 pp.

———. "The Western Sand Cherry." *South Dakota Agricultural Experiment Station Bulletin* 87 (June 1904), 64 pp.

Hargreaves, Mary W.M. "Hardy W. Campbell (1850–1937)." *Agricultural History* 32 (January 1958): 62–65.

Harris, Lionel. "Vegetable Variety Tests at the Scottsbluff Sub-Station." *Nebraska Agricultural Experiment Station Bulletin* 300 (June 1936), 27 pp.

Hatton, John H. "A Review of Early Tree-Planting Activities in the Plains Region." In *Possibilities of Shelterbelt Planting in the Plains Region*, ed. staff at Lake States Forest Experiment Station, 51–57. Washington, D.C.: Government Printing Office, 1935.

[Hays Agricultural Experiment Substation.] Annual Reports, 1912–1951 (typescripts). Hays: Western Kansas Agricultural Research Station.

Hedrick, U. P. *A History of Horticulture in America to 1860*. New York: Oxford University Press, 1950.

Hewitt, William L. "The 'Cowboyification' of Wyoming Agriculture." *Agricultural History* 76, no. 2 (Spring 2002): 481–494.

Hildreth, Aubrey C. "Climatic Factors Affecting Tree Growth on the High Plains." *Arborist's News* 18, no. 12 (December 1953): 105–114.

———. "Determination of Hardiness in Apple Varieties and the Relation of Some Factors to Cold Resistance." *Minnesota Agricultural Experiment Station Technical Bulletin* 42 (1926), 37 pp.

———. "Horticulture on the Wyoming Plains." In *Development of Horticulture on the Northern Great Plains*, ed. W. H. Alderman, 120–132. St. Paul: Great Plains Region, American Society for Horticultural Science, 1962.

———. "Ornamental Hedges for the Central Great Plains." *USDA Farmers' Bulletin* 2019 (1950), 25 pp.

Hildreth, A. C., and LeRoy Powers. "The Rocky Mountain Strawberry as a Source of Hardiness." *Proceedings of the American Society for Horticultural Science* 38 (1940): 410–412.

Howard, Gene S. "Herbaceous Perennials for the Central Great Plains." *USDA Agricultural Research Service Bulletin*, series 34, no. 71 (1965).

———. "Recommended Horticultural Plants Generally Hardy and Adaptable in the Central Great Plains Region." USDA Agricultural Research Service B-770 (February 1982; repr. September 1999), 6 pp.

———. "A Study of Shelterbelt Tree and Shrub Species under Dryland Culture in the Central Great Plains Region." M.S. thesis, University of Wyoming, 1961.

Howard, Gene S., and G. B. Brown. "Hardy, Productive Tree Fruits for the High Altitude Section of the Central Great Plains Region." *USDA Agricultural Research Service Bulletin*, series 34, no. 40 (1962), 6 pp.

———. "Seven Species of Broadleaf Deciduous Trees for Windbreaks: Effect of Spacing Distance and Age on Their Survival and Growth at Cheyenne, Wyoming." *USDA Technical Bulletin* 1291 (1963), 16 pp.

———. "Twenty-Eight Years of Testing Tree-Fruit Varieties at the Cheyenne Horticultural Field Station, Cheyenne, Wyoming." *USDA Agricultural Research Service Bulletin*, series 34, no. 39 (October 1962), 41 pp.

Jackson, Wes. *New Roots for Agriculture*. 2nd ed. Lincoln: University of Nebraska Press, 1985.

Jenkins, James F. "My Life Story" [1925]. Typescript from notes transcribed by his daughter, Agnes Metcalf. Jenkins Collection, Wyoming State Archives, Cheyenne.

Johnsgard, Paul A. *This Fragile Land: A Natural History of the Nebraska Sandhills*. Lincoln: University of Nebraska Press, 1995.

Johnson, E. W. "Hardy Trees and Shrubs for Western Kansas." *Kansas Agricultural Experiment Station Bulletin* 270 (1934), 32 pp.

Johnstone, Paul H. "In Praise of Husbandry." *Agricultural History* 11 (April 1937): 80–95.

Kellogg, Royal S. "Notes on the Native Woody Species of Western Kansas" (typescript). Hays: Western Kansas Agricultural Research Center [1903], 42 pp.

Kindscher, Kelly. *Edible Wild Plants of the Prairie: An Ethnobotanical Guide*. Lawrence: University of Kansas Press, 1987.

Kirkwood. William P. "The Romantic Story of a Scientist: The Work of Professor Hansen in Discovering and Inventing Fruits and Forage That Will Stand Sub-Zero Weather." *The World's Work* 15, no. 6 (April 1908): 10109–10120.

Kluger, James R. *Turning on Water with a Shovel: The Career of Elwood Mead*. Albuquerque: University of New Mexico Press, 1992.

Koster, George E. *A Story of Highway Development in Nebraska*. Lincoln: Nebraska Department of Roads, 1997.

Lana, E. P., ed. *A History of Horticulture of the Great Plains Region, 1960 through 1981*. St. Paul: North Central Region, American Society for Horticultural Science, 1984.

Lane, Frank P. "A History of Agricultural Extension Service in Wyoming Counties" (typescript). Laramie: Agricultural Extension Service, 1964.

Lind, Michael. "The New Continental Divide." *Atlantic Monthly* (January–February 2003): 86–88.

MacMechen, Edgar C. *Robert W. Speer, a City Builder*. Denver: Smith-Brooks, 1919.

McClelland, J. H., and Blanche E. Hyde. *History of the Extension Service of Colorado State College, 1912–1941*. Fort Collins: Extension Service, Colorado State College of Agriculture and Mechanic Arts, 1941.

Mikesell, W. A. "Forestry in Rawlins County." Kansas State Horticultural Society Annual Report 16 (1886): 128.

Miller, John E. "Eminent Horticulturalist: Niels Ebbesen Hansen." In *South Dakota Leaders: From Pierre Chouteau, Jr. to Oscar Howe*, ed. Herbert T. Hoover and Larry J. Zimmerman, 270–281. Vermillion: University of South Dakota Press, 1989.

Morrill, W. J. "Trees for Non-Irrigated Regions in Eastern Colorado." *Colorado Agricultural College Extension Service Bulletin* 123, Series 1 (September 1917): 3–20.

Muller, Edward K., ed. *DeVoto's West: History, Conservation, and the Public Good*. Athens, Ohio: Swallow, 2005.

Nelson, Aven. "The Grain Smuts and Potato Scab." *Wyoming Agricultural Experiment Station Bulletin* 21 (1895): 5–24.

———. "Native Vines in Wyoming Homes." *Wyoming Agricultural Experiment Station Bulletin* 50 (March 1902): 1–15.

———. Papers. University of Wyoming American Heritage Center, Laramie.

———. "The Russian-thistle." *Wyoming Agricultural Experiment Station Press Bulletin* 1, reprinted in Wyoming Agricultural Experiment Station Annual Report (1895): 29–31.

———. "The Trees of Wyoming and How to Know Them." *Wyoming Agricultural Experiment Station Bulletin* 40 (January 1899): 59–110.

———. "The Winter-Killing of Trees and Shrubs." *Wyoming Agricultural Experiment Station Bulletin* 15 (December 1893): 213–222.

———. "The Worst Weeds of Wyoming and Suggested Weed Legislation." *Wyoming Agricultural Experiment Station Bulletin* 13 (1896): 267–320.

Nelson, Elias. "The Shrubs of Wyoming." *Wyoming Agricultural Experiment Station Bulletin* 54 (July 1902): 1–47.

Olson, James C. *J. Sterling Morton, Founder of Arbor Day*. 2nd ed. Lincoln: Nebraska State Historical Society Foundation, 1972 [1942].

Opie, John. *Ogallala: Water for a Dry Land.* 2nd ed. Lincoln: University of Nebraska Press, 2000 [1993].

Overfield, Richard A. *Science with Practice: Charles E. Bessey and the Maturing of American Botany.* Ames: Iowa State University Press, 1993.

Pabor, William E. *Colorado as an Agricultural State.* New York: Orange Judd, 1883.

Parkman, Francis. *The Oregon Trail.* New York: Literary Classics of the United States, 1991 [1849].

Payne, James E. "Investigation of the Great Plains. Field Notes from Trips to Eastern Colorado." *Colorado Agricultural Experiment Station Bulletin* 59 (1900), 19 pp.

———. "Notes on a Dry Land Orchard." *Colorado Agricultural Experiment Station Bulletin* 173 (1910), 7 pp.

———. "Unirrigated Lands of Eastern Colorado: Based on a Study and Residence of Seven Years." *Colorado Agricultural Experiment Station Bulletin* 77 (1903), 16 pp.

Peper, Paula J., E. Gregory McPherson, James R. Simpson, Scott E. Maco, and Qingfu Xiao. "City of Cheyenne, Wyoming Municipal Tree Resource Analysis." Prepared by the Center for Urban Forest Research, USDA Forest Service, Pacific Southwest Research Station, Davis, California (2004), 62 pp.

Peters, Scott J., and Paul A. Morgan. "The Country Life Commission: Reconsidering a Milestone in American Agricultural History." *Agricultural History* 78, no. 3 (Summer 2004): 289–316.

Phillips, William M. "A History of the Agricultural Research Center–Hays: The First 100 Years." *Kansas Agricultural Experiment Station Bulletin* 663 (April 2001), 132 pp.

Pisani, Donald J. "Reclamation and Social Engineering in the Progressive Era." *Agricultural History* 57 (January 1983): 46–63.

———. *To Reclaim a Divided West: Water, Law, and Public Policy, 1848–1902.* Albuquerque: University of New Mexico Press, 1992.

Pool, Raymond J. "Fifty Years on the Nebraska National Forest." *Nebraska History* 34 (September 1953): 139–179.

Popper, Deborah E., and Frank Popper. "The Great Plains: From Dust to Dust." *Planning Magazine* 53, no. 12 (December 1987): 12–18.

Powell, John Wesley. *Report on the Lands of the Arid Region of the United States*, ed. Wallace Stegner. Cambridge: Harvard University Press, 1962 [1878].

Powers, LeRoy. "Early Cheyenne Pie Pumpkin." *USDA Circular* 537 (November 1939).

Preston, R. J., and J. F. Brandon. "37 Years of Windbreak Planting at Akron, CO." *Colorado Agricultural Experiment Station Bulletin* 492 (1946), 25 pp.

Quayle, W. L. "Trees—Wyoming's 25-Year Record." *Wyoming Agricultural Experiment Station Circular* 46 (September 1951), 11 pp.

Rasmussen, Wayne D. *Taking the University to the People: Seventy-Five Years of Cooperative Extension.* Ames: Iowa State University Press, 1989.

Robertson, John S. "Fruit Culture 1923 in the Southern Black Hills." *South Dakota Horticultural Society Annual Report* 21 (1924): 115–118.

Rydberg, Per A. "Flora of the Sand Hills of Nebraska." USDA Division of Botany, *Contributions from the U.S. National Herbarium* 3, no. 3 (September 14, 1895): 133–203.

Salac, Sotero S. "Collection, Propagation, Culture, Evaluation, and Maintenance of Plant Materials for Highway Improvement." Research report prepared by the Department of Horticulture, University of Nebraska. Lincoln: Nebraska Department of Roads, 1977, 185 pp.

Sandoz, Mari. *Old Jules: Portrait of a Pioneer.* New York: MJF Books, 1996 [1935].

Schlichter, J. B. "The Importance of Horticulture to a Successful Settlement of Western Kansas." *Kansas State Horticultural Society Proceedings* (1886): 51–54.

Scott, Charles A. "Trees for Kansas." *Kansas Agricultural Experiment Station Circular* 55 (January 1916), 19 pp.

Smith, Henry Nash. "Rain Follows the Plow: The Notion of Increased Rainfall for the Great Plains, 1844–1880." *Huntington Library Quarterly* 10 (1947): 181–185.

Smith, Shane. "The Hildreth/Howard Arboretum." White Paper. Cheyenne: Cheyenne Botanic Gardens, 2001.

Smythe, William E. *The Conquest of Arid America,* intro. Lawrence B. Lee. Seattle: University of Washington Press, 1969 [1899].

———. "Real Utopias in the Arid West." *Atlantic Monthly* 79 (May 1897): 599–609.

Steinel, Alvin T. *History of Agriculture in Colorado.* Fort Collins: Colorado State Agricultural College, 1926.

Stewart, Miller J. "To Plow, to Sow, to Reap, to Mow: The US Army Agriculture Program." *Nebraska History* 63 (Summer 1982): 194–215.

Street, W. D. "Vegetable Gardening in Northwest Kansas." Kansas State Horticultural Society Annual Report 26 (1896): 60–63.

Tice, John H. *Over the Plains, on the Mountains; or, Kansas, Colorado, and the Rocky Mountains: Agriculturally, Mineralogically and Aesthetically Described.* St. Louis: "Industrial Age" Printing, 1872.

Trelease, Frank J. "The Concept of Reasonable Beneficial Use in the Law of Surface Streams." *Wyoming Law Journal* 12 (Fall 1957): 1–21.

Vaplon, E. E. "Reports and Plans of Town and City Garden Clubs." *Colorado Agricultural Extension Service Bulletin* 135, Series 1 (February 1918), 15 pp.

Wallace, Henry A. *New Frontiers.* New York: Reynal and Hitchcock, 1934.

Ware. E. R., and Lloyd Smith. "Woodlands of Kansas." *Kansas State Agricultural Experiment Station Bulletin* 285 (July 1939), 41 pp.

Warren, Francis E. Papers. Boxes 21–22. University of Wyoming American Heritage Center, Laramie.

Watrous, Ansel. *History of Larimer County, Colorado.* Fort Collins: Courier, 1911.

Webb, Walter Prescott. *The Great Plains.* Lincoln: University of Nebraska Press, 1981 [1931].

———. "The Story of Some Prairie Inventions." *Nebraska History* 34, no. 4 (December 1953): 229–243.

Wedel, Waldo R. "Notes on the Prairie Turnip (Psoralea esculenta) among the Plains Indians." *Nebraska History* 59, no. 2 (Summer 1978): 154–179.

Werner, H. O. "Varieties of Tomatoes Recommended for Various Localities in Nebraska on the Basis of Their Physiological Adaptations." Nebraska Horticultural Society Annual Report (1938): 27–29.

Wilber, Charles Dana. *The Great Valleys and Prairies of Nebraska and the Northwest.* 3rd ed. Omaha: Daily Republican, 1881.

Williams, Roger L. *Aven Nelson of Wyoming.* Boulder: Colorado Associated University Press, 1984.

Willard, James F., ed. *The Union Colony at Greeley, Colorado, 1869–1871.* Boulder: University of Colorado Historical Collections, 1918.

Wilson, Edward O. *Biophilia.* Cambridge: Harvard University Press, 1984.

Wilson, William H. *The City Beautiful Movement.* Baltimore: Johns Hopkins University Press, 1989.

Wood, Asa B., ed. *Pioneer Tales of the North Platte Valley and the Nebraska Panhandle.* Gering, Neb.: Courier, 1938.

Woodward, Harry R. "A Great Fruit Grower and His Contribution [John Robertson]" (typescript). Brookings: South Dakota State University, 1941.

Index

Page numbers in italics indicate illustrations

Agriculture, compared to horticulture, 3. *See also* Farming, dry-land; Irrigation; Jeffersonian idealism; U.S. Department of Agriculture
Ainsworth, Nebraska, 70, 73
Akron, Colorado, dry-land experiment station, 118, 177, 181, 218
Alfalfa, imported from Asia, 153, 156; as soil restorative, 38
Allen, Martin, 27
Alliance, Nebraska, 236
Altman, Henry, 48
Apples, 15, 203, 206; Anoka (cultivar), 154; Ben Davis (cultivar), 28, 146; in Colorado, 28, 40, 115, 117, 146; Duchess of Oldenburg (cultivar), 146; fire-blight, 207; Haralson (cultivar), 205; in Kansas, 27, 122; in Nebraska, 20, 68, 111; Red Astrakhan (cultivar), 146; in South Dakota, 153–154; Wealthy (cultivar), *100*, 146, 205; in Wyoming, 89, *100*
Apricot, 115, 122
Aquifer. *See* Ogallala aquifer
Arbor Day: in Colorado, 130, 133–134, 142; founding of, 13; in Nebraska, 19, 23, 185; Wyoming, 17. *See also* Tree City, USA
Arboretum. *See* Cheyenne Horticultural Field Station; Nebraska
Archer, Wyoming, agricultural experiment station, 197, 198
Ardmore, South Dakota, USDA field station, 219
Arid lands, classification of, 55
Arkansas River, 2, 10; valley, 15, 28, 107, 147
Artichoke, Jerusalem, 121, 209

Ash, 120, 124; green, in Kansas, 27, 28, 123, 124; green, in Nebraska, 185; white, in Colorado, 146
Asparagus, 12, 27; in Colorado, 42; in Kansas, 120; in Wyoming, 208, *213*, 214
Atwood, Kansas, 120, 235
Aughey, Samuel, 26, 52–53, 69

Babb, Myron F., 204, 209, 210, 214, 217
Bailey, Liberty Hyde, 10, 174. *See also* Country Life Commission
Balm-of-gilead, in Wyoming, 88
Barberry, in Wyoming, 88
Barbour, Erwin H., 44, 112
Barteldes and Company, 146
Bassett, S. C., 81
Bates, J. M., 70
Beans, 8; in Colorado, 114; in Kansas, 120, 121; in Nebraska, 187; in Wyoming, 88, 209, 217
Beautification, community, 4, 86, 196, 243; homestead, 130, 217; roadside, 235–236
Beetles. *See* Insects
Beets: in Kansas, 120, 121; in Nebraska, 188; in Wyoming, 46, 89, 208, *214*, 217
Bellvue, Colorado, 15, 158
Bent's Fort, Colorado, 10
Berthoud, Colorado, 43
Bessey, Charles, 2, 3, 10, 60, 115, 133, 174, 244; and Aven Nelson, 86, 90, 96; applying science to horticulture, 64–65, 71; education, 65; and Hatch Act, 64–65; interest in trees, 69, 72–73; on limits of horticulture on the High Plains, 78–79; and Nebraska National Forest, 19, 73, 74–75; and Niels Hansen, 152, 157, 159, 161; professor of botany and horticulture, 65–66; public service by, 68, 69, 70, 81; state botanist, 68, 71, 80; and State Horticultural Society, 68, 80–81; as teacher, 69–70; on usefulness of liberal arts to prospective horticulturists, 65–67. *See also* U.S. Forest Service, Bessey Nursery
Big Creek, Hays, Kansas, 121
Big Thompson River, Colorado, 14, 15; valley of, 2, 8, 59
Binkley, A. M., 186
Biophilia, 244
Birch, in Denver, 145
Bird City, Kansas, *246*, 246–247

Blackberries: in Colorado, 146; in Kansas, 122
Black Hills, South Dakota, 76, 78
Blount, A.E., 59
Boomers, 26, 52, 79, 108. *See also* Aughey, Samuel; Elliott, Richard S.; Wilber, Charles Dana
Botanic gardens. *See* Cheyenne; Denver
Botany: applications to horticulture, 65–66, 67–68; compared to horticulture, 152. *See also* Science, applications of
Boulder Creek, Colorado 14
Boxelder: in Colorado, 42; in Kansas, 27, 28, 120; in Nebraska, 73, 85
Boyd, David, 34, 59
Brandon, J. F., 124
Brimmer, George E., 196–199, 201–202, 203
Brooks, Bryant B., 97, 174
Brule, Nebraska, 235
Bruner, Lawrence, 73
Buckthorn, 217
Budd, Joseph L., 152, 159
Buffaloberry, 13, 70; in Wyoming, 88
Buffum, Burt C., 87–89
Burpee Company, W. Atlee, 146
Byers, William N., 14, 37, 42

Cabbages, 14; in Colorado, 30; in Kansas, 120, 121; in Nebraska, 79, 187; in Wyoming, 30
Cache la Poudre River, Colorado, 14, 15, 33, 38, 39, 158; conflict over, by Greeley and Fort Collins irrigators, 41, 59; valley of, 28
Cambridge, Nebraska, 236
Campbell, Hardy W., 109–111, 118
Cameron, Robert A., 36, 38
Canals. *See* Irrigation
Cañon City, Colorado, 28
Cantaloupes, Rocky Ford, 147
Caragana, 11; in Wyoming, 183, 217
Carey Act, 89, 171
Carey, Joseph M., 2, 46, 88–89, 171–172
Carrots: in Kansas, 120, 121; in Nebraska, 187; in Wyoming, 89, 208, 217
Cassidy, James, 58, 81, 147
Catalpa, in Kansas, 123, 124
Cauliflower: in Nebraska, 187; in Wyoming, 208, 214, *215*
Cedar, red: in Kansas, 122, 124, 181; in Nebraska, 78, 235; in Wyoming, 219. *See also* Juniper

Celery: in Colorado, 30; in Nebraska, 187
Central High Plains Field Station. *See* Cheyenne Horticultural Field Station
CES. *See* Cooperative Extension Service
Chard, Swiss: in Nebraska, 188; in Wyoming, 208, 214
Cherries, 15; black, 73; black, in Colorado, 115, 117; black, in Kansas, 122, *123*; black, in Nebraska, 111; black, in Wyoming, 89, 206; sand, 10, 13, *158*, 207; sand, in South Dakota, 157–161; sour, in Colorado, 40; in Kansas, 122. *See also* Chokecherry
Cherry Creek, Colorado, 42, 143. *See also* Denver
Cheyenne, Wyoming, 35, 47–48, 174, 195, 198; botanic gardens, 223, 234; climate at, 196; Industrial Club in, 201; USDA agricultural experiment station, 111, 197. *See also* Cheyenne Horticultural Field Station
Cheyenne Horticultural Field Station, viii, 4, 10, 77, 145, 165, 166, 167, 191, 200, *201, 202*; arboretum, viii, 222–223; fruit growing at, *164*, 205–208; irrigation for, 199, 203; mission of, 195, 198; ornamentals at, 217–218, *218*; physical facilities at, 199; reason-for-being of, 196–197; research workplan of, 203–204; transition to grasslands research at, 220–222, *222*; trees and shrubs at, 201, 218–220, *221*; vegetables at, 208–215, *212*, 217
Cheyenne River, South Dakota, 160
Cheyenne Wells, Colorado, 107, 113. *See also* Colorado State University
Chilcott, E. C., 118, 166, 198, 199, 201
Chives, 208
Chlorosis, 220, *221*, 231
Chokecherry, 42, 70; in Wyoming, 88, 217
Chrysanthemum, 145, 195, 217, 218
City Beautiful movement, 144–145. *See also* Denver
Claims clubs, 15–16
Clark, J. Max, 40
Clarke-McNary Act, 180–182, 183, 184, 219
Clear Creek, Colorado, 14, 15; valley of, 28
Clematis, western, 95
Coffeetree, Kentucky, in Kansas, 123
Coffin v. Left Hand Ditch Co., 58
Colby, Kansas, 218, 237
Cole, Mrs. M.D., 43–44

Collards, 214
Colorado: constitution of, 31, 58; Seed Act of, 178–179; State Board of Agriculture, 43; state forester, 180, State Horticultural Society, 22, 29–30, 43, 132, 146; state horticulturist of, 179; tree planting in, 31, 33. *See also* Colorado State University; Northern Colorado Horticultural Society
Colorado Agricultural College. *See* Colorado State University
Colorado Nursery Company, 146
Colorado Seed House. *See* Barteldes and Company
Colorado State University, Fort Collins, 30, 59, 87, 88, 181; agricultural experiment station, 59, 147, 177; Cheyenne Wells substation, 107; Cooperative Extension Service, 176, 228, 230–231, 232, 233; county agents through, 173, 176, 177; experimental orchard at, 179; seed distribution and testing by, 43, 146, 178–179, 233
Communities, 5, 239; creating permanent, viii; livable, ix
Community foundation: Bird City, Kansas, 247; Nebraska, 246; Wyoming, viii, 55
Condra, George E., 185
Conine, Martha A.B., 139
Cooperative Extension Service, 174–175, 177, 183, 186. *See also individual universities by name*
Corn, sweet, 8; in Kansas, 121; in Nebraska, 187; in Wyoming, 11, 89, 217
Cotoneaster, 217
Cottonwoods, in Colorado, 31, 42, 145; in Kansas, 27, 120, 181; in Nebraska, 185; in Wyoming, 47, 88, 89, 136
Country Life Commission, 4, 173–174
County agents, 165, 183, 184, 219
Crabapples: Dolgo or Siberian, 154, 156, 206; in Kansas, 122; native prairie, 154; in Wyoming, 206
Crandall, Charles S., 81, 147, 148
Crane, Arthur G., 198, 199
Crow Creek, Wyoming, 113, 204
Crowley, J.H., 28
Cucumber, cultivated: in Colorado 30, 114; in Kansas, 120; in Nebraska, 79, 187; wild, 95; in Wyoming, 95, 208, 217
Cultivar. *See* Plants: nomenclature
Curley, Edwin A., 24–25

Index 263

Currants, 13; in Colorado, 42, 115, 146; in Wyoming, 88, 89, 207

Dakota Territory, 59, 86
DeBoer, Saco Rienk, 136–137, 199, *141, 142*
Delzell, James E., 81
Deming, W. C., 184
Denver, Colorado, 227–229; Board of Water Commissioners (Denver Water), 140, 228–230; botanic gardens, 220, 232; Capitol Hill neighborhood, 138, 139; as city beautiful, 138–145, *141, 144*; as consumer of locally grown produce, 137–138; Lower Downtown (LoDo), 227; parks in, 130, 139, 140, *142, 143,* 145, 218; Society of Ornamental Horticulture, 145. *See also* Cherry Creek, Colorado; Xeriscaping
DeVoto, Bernard, 53, 54, 172
Dismal River, Nebraska. *See* U.S. Forest Service
Downing, Andrew Jackson, 243–244
Drought, 27, 44, 79, 123, 238; dust storms as result of, 114, 186; normal condition of, 54
Dubois, William, 199

Eastwood, Alice, 133
Eastwood, J. W., 147
Eaton, Benjamin, 130
Eckert, Peter, 117
Edmondson, William O., 177, 183–184, 187
Eggplant, in Nebraska, 187
Elliott, Richard S., 26–27, 52
Ellis, Kansas, 26
Ellsworth, Nebraska, 204
Elm, 143, 145; American, in Kansas, 181; American, in Nebraska, 185; Chinese, in Kansas, 181; Chinese, in Wyoming, 187; red, in Kansas, 28, 123; Siberian, 219; white, in Kansas, 123, 124
Emerson, John L., 182, 204, 219
Endive, in Nebraska, 188
Engstrom, Harold F., 204
Enlightenment, legacy of the, ix, 16, 22, 133. *See also* Jeffersonian idealism
Experiment stations, agricultural, 3, 64. *See also individual locations by name*
Eyerly, Gene, 228

Fairs: county, 134; state, 71, 98, 99
Farmers' Institutes, 40–41, 69, 98, 165

Farmers markets, 234
Farming, dry-land: congresses about, 110, 111; defined, 108; limits of, 113; principles of, 108–109. *See also* Xeriscaping
Farms: commodity, viii; demonstration, 173; stock, 4, 46, 55
Farmsteads, 25
Fatzer, Harold R., 181
Fernow, Bernhard E., 73
Ferry Company, D. M., 146–147, 177
Feucht, James R., 228, 230–232
Field and Farm (newspaper), 21; on Arbor Day, 133–134; on county fairs, 134; on "easternizing," 129–131; against jackrabbits, 125, 126; opposition to Congressional seed distribution, 147; on ornamental horticulture, 31, 145
Flagler, Colorado, 116, 117
Flowers, ornamental, 7, 13, 47, 146, 189, 217, *218, 239. See also* Chrysanthemum
Ford, R. E., 182
Fort Collins, Colorado, 233–234, 245
Fort D.A. Russell, Cheyenne, Wyoming, 47, 48. *See also* Fort F. E. Warren
Fort F. E. Warren, Cheyenne, Wyoming, 195
Fort Laramie, Wyoming, 10–12, 25; Fort Leavenworth as supply depot for, 12, 25; hospital flower garden at, 47, 244
Fort William, Wyoming. *See* Fort Laramie, Wyoming
Fourier, Charles. *See* Utopianism
Front Range (Colorado), 4, 14, 15, 16. *See also individual towns and waterways by name*
Frazier, Jesse, 28
Fruit growing, 14; in Colorado, 28–29, 33, 116, 146, 179; dry-land, 110; dry-land compared to irrigated, 206–207; in Kansas, 122, *123;* in Nebraska, 20, 111, *112,* 180; in South Dakota, 165; in Wyoming, 86, 89, 97, *206. See also* Cheyenne Horticultural Field Station, fruit growing; *individual varieties by name*
Furnas, Robert W., 13, 20, 69

Garden City, Kansas, 109, 118, 237
Garden clubs: in Greeley, Colorado, 176; in Nebraska, 189
Gardens, ix, 224; classification of, 175–176; irrigated, on farmstead, 44, 54, 113, 117;

irrigated, in towns, 42; manuals about, 186–187, 214, 217; market, 14, 15, 137; ornamental, 4, 46–47; at Shadyside (Denver), 42–43; school, 130–132; urban, 175, *176,* 189, 210; vegetable, 7, *11,* 12, 46, 112, 120. *See also* Cheyenne; Denver; *individual plants by name*
Garlic, 188
Gauger, J. E., 147
Gering, Nebraska, 136, 236
Gipson, Arthur E., 30, 146, 159
Gipson, Mrs. Arthur E., 132
Godi, Donald H., 228
Golden, Colorado, 15
Goodland, Kansas, 247
Goodwin, M. J., 90
Gooseberries, 27; in Colorado, 115, 117, 146; in Wyoming, 88, 89, 207. *See also* Currants
Gordon, John H., 90, 96, 112
Grabow, Jim, 228
Grand Island, Nebraska, 2, 22
Grapes, bush, 13; in Colorado, 117, 146; in Kansas, 27, 122; in Nebraska, 20; riverbank, 95; in Wyoming, 112
Grasses, buffalo, 190–191, 230, 238–239; blue grama, 72, 190
Grasshoppers. *See* Insects
Great American Desert, 1, 21, 107
Great Plains, The (Webb), 2
Greeley, Colorado, 3, 33–42, *37,* 56, 89–90; as "Garden City," 134–135, *135;* vegetable gardens in, 175–176. *See also* Union Colony of Colorado
Greeley, Horace, 34, 36
Greeley Nursery, *41,* 146
Green industry, 4, 223, 231, 232, 244
Gregg, Josiah, 52
Grenfell, Helen, 131–133, 232
Groundwater, 4, 237–239. *See also* Water
Gurney's Nursery, 151, 166, 205

Hackberry: in Colorado, 118; in Kansas, 28, 123, 181; in Nebraska, 73
Hail, damage from, 89, 114, 196, 208, 210
Hall, William L, 74–75
Halsey, Nebraska. *See* U.S. Forest Service
Haney, J. G., 121–122, 124
Hansen, Clifford P., 222
Hansen, Niels E., 4, 64, 81, 115, 148, 207, 233, 239; bridging science and practice, 165–166; and Charles Bessey, 152, 157, 159, 161; education, 152–153; and Luther Burbank, 152, 157, 161; as plant breeder, 152, 154–156, *162,* 166; as plant explorer, 152; sand cherries, experiments with, 157–161; and South Dakota Horticultural Society, 151, 165
Hansen Nursery, 166
Hardiness, of plants, 28, 154, 156, 166, 203, 217; definition of, 153
Harrison, Charles S., 8, 20–22, 27, 244
Hartley, Carl, 77
Hatch Act, 3, 64–65, 68, 88, 107
Hay Springs, Nebraska, 111
Hays, Kansas, 2, 27. *See also* Kansas State University
High Plains, *vi;* geography of, 2
Hildreth, Aubrey C., 167, 202–205, 208, 217, 220
Hill City, Kansas. *See* Pomeroy Model Farm
Hitchcock, Phineas W., 19
Home beautiful, 37, 85, 95, 134, 165
Homestead Act, 16, 46, 54
Hops, common, 95
Horse Creek, Wyoming, 90
Horseradish, 27; in Kansas, 120
Horticultural Field Station. *See* Cheyenne Horticultural Field Station
Horticulture: civilizing role of, vii, 23, 33, 36, 44, 93; and community, 5, 86, 96, 233–234; definition of, 2–3; as expression of civic-mindedness, 33, 243–244; as expression of sentimentality, ix; keeping families together, 245; as means of moral education, 132; for newcomers to High Plains, 231–232; ornamental, 13, 42, 95, 137, 145, 146; preventing emigration of young people, 134; as profession, 233; as secular religion, viii; therapeutic, 244; and women, 7, 8, 12, 43–44, 136, 139, 175, 187, 244; and youth, 176
Horticulturists, proposal for county, 165–166
Hot Springs, South Dakota, 165
Howard, Gene S., 218, 220
Howell, James, 116–117
Hoyt, John W., 46
Hyannis, Nebraska, 76

Imperial, Nebraska, 245
Indians, Plains, ix, 8, 11, 160

Index 265

Insects: beetles, blister, 114, 210; beetles, potato, 114; borers, black locust, 137; grasshoppers, 27, 114, 210; insecticides against, 114, 147; moths, coddling, 207
Iowa State College (University), 64, 152, 159
Irrigation: on benchlands, 38; canals and ditches for, 14, 33, 39, 41, 42, 203; contributing to community permanence, 136; definition of, 55; districts, 56, 57–58; early years of, 12, 14, 33, 38, 78; flumes, 42, 120; headgates, 14, 42, 120, 203. *See also* Cheyenne Horticultural Field Station; Mead, Elwood

Jackrabbits, 124–126, *125, 126*
Jackson, Wes, 239
Jardine, William M., 196, 197, 198, 199
Jeffersonian idealism, ix, 4, 54, 56, 152, 243. *See also* Enlightenment
Jenkins, James F., 47, 89, 136
Johnsgard, Paul, 245
Johnson, E. W., 181, 182
Johnson, Fred R., 182
Johnston, Martin R., 89
Julian, Frank, 101–102
Juniper, in Colorado, 145. *See also* Cedar

Kalaidis, Panayoti, 232
Kale, 208
Kansas: legislature, 25, 59, 121; State Agricultural Society, 27; state forester, 181; State Horticultural Society, 22, 26, 237; tree planting in, 25–27. *See also* Groundwater; Kansas State University
Kansas Agricultural College. *See* Kansas State University
Kansas Farmer (newspaper), 26, 27
Kansas Nursery and Landscape Association, 237
Kansas State University, 114, 121, 124; agricultural experiment station, 28; Cooperative Extension Service, 237; Division of Forestry, 124. *See also* Western Kansas Agricultural Research Center
Kearney, Nebraska, 75, 236
Keesen, Larry, 228
Kellogg, Royal S., 75, 123
Kessler, George E., 145
Kinkaid Act, 80, 184
Klett, James E., 232

Knapp, Seaman A., 64, 65, 152, 173
Knight, Henry G., 100
Kohlrabi, in Nebraska, 187
Kohankie, Adam, 145
Krause, James E., 204

Lamb's quarter, 12
Land-grant colleges. *See* Universities, land-grant
Laramie, Wyoming, 234. *See also* Wyoming, University of
Laramie River, Wyoming, 12, 47, 88
Leavy, John, 135
Lee, Henry, 15
Leek, in Nebraska, 188
Lettuce, 14; in Colorado, 114; in Kansas, 120; in Nebraska, 188; in Wyoming, 217
Lilacs, 145, 217
Limon, Colorado, 107, 182
Lind, Michael, 247
Livestock, protection from, 24, 39, 48
Locust: black, in Colorado, 42, 116, 181; black, in Kansas, 120; black, in Nebraska, 73; honey, in Colorado, 181; honey, in Kansas, 27, 123, 124; honey, in Nebraska, 185
Lodgepole Creek, Nebraska, 78
Lone Pine Creek, Nebraska, 70
Long, Stephen, 1
Longmont, Colorado, 131
Loup River, Nebraska, 44. *See also* U.S. Forest Service
Loveland, Colorado, 15
Loveland Nursery, 205
Lucerne. *See* Alfalfa
Lupton's Fort, Colorado, 10

Mandan, North Dakota. *See* Northern Great Plains Field Station
Maples: in Denver, 115; in Kansas, 123
Master Gardeners, 232, 234, 237
McClelland, C. K., 123
McClelland, James S., 30
McGee, Gale W., 222
McMullen, Adam, 185
Mead, Elwood, 14, 58, 59–60, 108, 171
Meeker, Nathan C., 33–41, 51, 145
Melons, 14, 114; in Kansas, 120; in Nebraska, 187; in Wyoming, 89, *213*. *See also* Cantalopes; Watermelons

266   Index

Mickelson, George T., 167
Miller, Bill, 228
Mikesell, W.A., 120
Moonlight, Thomas, 46
Moorehead, John H., 81
Morrill Act, 17
Morrill, W. J., 181
Morton, J. Sterling, 13, 19, 20; "Fruit Address" by, 22–23
Mulberry, in Colorado, 115; in Kansas, 28, 123
My Antonia (Cather), 13

Nebraska, 65, 93; Academy of Sciences, 52; Department of Roads, 236; legislature, 13, 25, 59; State Board of Agriculture, 20, 69, 79; state fair, 71; State Horticultural Society, 20, 21, 22, 66, 68, 69, 71, 75, 80, 111; statewide arboretum, 236; tree planting in, 19–22, 21, 185, 235. *See also* Arbor Day; Nebraska, University of
Nebraska Farmer (newspaper), 68
Nebraska Forest Reserves. *See* U.S. Forest Service
Nebraska Nursery and Landscape Association, 236
Nebraska, University of, 44, 52, 64, 65, 69; agricultural experiment station, 69, 111, 187–188; Conservation and Survey Division, 185; Cooperative Extension Service, 188, 236; herbarium and horticulture department, 66; state forester, 185
Nelson, A. L., 197
Nelson, Aven, 2, 3, 64, 111, 151, 174, 204, 244; applying botany to horticulture, 87, 90, 92; and Charles Bessey, 81, 86, 90, 96; education, 87; horticultural publications, 86, 92–93, 97; on "making a home," 85–86, 94; on moral virtue of horticulture, 86, 91; professor of botany and horticulture, 87; as secretary to State Horticultural Society, 100–102; as secretary to State Board of Horticulture, 96–97, 99–100; as state botanist, 87; as teacher, 86, 93
Nelson, Elias, 94
New Deal, 75, 185, 186, 191; Civilian Conservation Corps of, 195, 199, 204; Prairie States Shelter Belt Project, 185
Newell, Frederick H., 75
Newlands, Francis G., 172
Niobrara River, Nebraska, 70

North Platte, Nebraska, 75, 236; USDA agricultural experiment station, 219
Northern Colorado Horticultural Society, 29–31
Northern Colorado Nursery, Fort Collins, 205
Northern Great Plains Field Station, Mandan, North Dakota, 180, 183, 196, 201, 219
Northwest Kansas Groundwater Conservation Foundation, 238. *See also* Groundwater
Nurseries, 28, 30, 166, 208, 219, 232; in Colorado, 41, 145–146, 205; in Kansas, 122–124; in Nebraska, 111; in South Dakota, 151, 166, 205; in Wyoming, 97–98, 102. *See also* U.S. Forest Service, Bessey Nursery; *individual nurseries by name*
Nye, Bill, 1, 86, 101

Oak: bur, in Kansas, 123, 181, 327; yellow chestnut, in Nebraska, 236
Oberlin, Kansas, 120
Ogallala, Nebraska, 245
Ogallala aquifer, 237–238
Olive, Russian, 24, 42; in Colorado, 115; 145; in Kansas, 181; in Wyoming, 187
Olmsted, Frederick L., Jr., 145
Onions, 12, 14; in Colorado, 114; in Kansas, 120; in Nebraska, 79, 188; in Wyoming, 46, 88, 208, 217
O Pioneers! (Cather), 13
Orach, 208, 210
Orchards. *See* Fruit growing
Origin of Cultivated Plants (de Candolle), 154
Orman, James B., 140
Osage-orange, 24, 123
Owens, Bill, 238

Pabor, William E., 42–43
Parker, Colorado, 110
Parkman, Francis, 1
Parks, municipal, 40, 246; in Cheyenne, 47–48, 99. *See also* Denver
Parsley: in Nebraska, 187; in Wyoming, 217
Parsnip: in Kansas, 121; in Nebraska, 187; in Wyoming, 208
Parsons, E. R., 110
Pathfinder Dam, Wyoming, 172
Payne, James E., 3, 107, 112, 114–118, 173

Peaches, 15; in Colorado, 117; in Kansas, 27, 122; in Nebraska, 20
Pears, 15; in Colorado, 31, 146; in Kansas, 27, 122; in Nebraska, 20; in Wyoming, 206, 207
Peas, 14; in Colorado, 114; in Kansas, 120; in Nebraska, 187; in Wyoming, 46, 88, 89, 217
Pennock, Charles E., 147–148, 158–159
Peppers, in Nebraska, 187
Persimmons, in Kansas, 123
Pierre, South Dakota, 2
Pike, Zebulon, 1
Pinchot, Gifford, 75
Pine, 145; Austrian, in Kansas, 73, 122; Austrian, in Nebraska, 185; jack, 73, 78; jack, in Kansas, 181; jack, in Nebraska, 185; lodgepole, 93; ponderosa, in Colorado, 118, *119*; ponderosa, in Kansas, 122, 181, 182; ponderosa, in Nebraska, 70, 73–74, 76, 78, 185; ponderosa, in Wyoming, 183, 219; Scotch, in Kansas, 122; Scotch, in Nebraska, 73
Pingree Gardens, 139. *See also* Victory Gardens
Plants, breeding of, 147, 155–156, 211, *213, 215*; cold frames, 209, *213*; "damping off," 77; grafting, 29, 120, 146, 160–161, 206; GreatPlants (Nebraska), 236–237; "hardening off," 188, 209; hotbeds, 188, 209; nomenclature of, 8–10, 93–94, 153; PlantSelect (Colorado), 232, 233; Prairie State Flowers (Kansas), 237; Pride of Kansas, 237; seedbeds, *216. See also* Weeds
Plants, native, 10, 12, 47, 69; cultivation of, 88, 94; water requirements, 177. *See also* Grasses; Plants; Vines
Platte River (Colorado, Nebraska, Wyoming), 44; South, 14, 41, 59; North, 12, 78, 172
Plums, 12, 15, 160; in Colorado, 115, 117, 145, 148; in Kansas, 27, 122; in Nebraska, 20, 68, 111; Opata (cultivar), 161, 163, *164;* Sandoz (cultivar), 111; Sapa (cultivar), 161, 163; in South Dakota, 161, 163; in Wyoming, 89, 207
Pomeroy Model Farm, 110
Pool, Raymond, 76
Poplars, Carolina or Norway: in Colorado, 143, 145; in Kansas, 123; in Nebraska, 185

Popper, Deborah, 245
Popular Flora of Denver (Eastwood), 133
Potatoes, 10, 12; in Colorado, 113, 114, 177; early Ohio (cultivar), 113, 120; early rose (cultivar), 114; Greeley spud (cultivar), 38–39; in Kansas, 120, 121; in Nebraska, 79, 187; scab on, 90–91; in Wyoming, 46, 88, 89, 113
Powell, John Wesley, 3, 48, 51, 53–58, 60, 64, 171, 172, 239
Powers, LeRoy, 204, 208, 220
Preston, Douglas A., 102
Privet, Cheyenne, 217
Pumpkin, in Colorado, 114; early Cheyenne pie (cultivar), 210–211, *212*; in Kansas, 120; in Nebraska, 187; in Wyoming 89, 208, 217

Quayle, W. T., 197

Radishes, 10; in Colorado, 14, 114; in Kansas, 120, 121; in Nebraska, 187; in Wyoming, 217
Railroad, 3, 12; Burlington and Missouri, 20, 109; Denver, South Park, and Pacific, 137; Denver, Utah, and Pacific, 137; Kansas Pacific, 26–27, 113; Union Pacific, 47, 109, 137
"Rain follows the plow," 52–53, 108, 114
Ranching, viii, 79–80, 115, 117; as agriculture, 75. *See also* Livestock; Wyoming Stock Growers Association
Raspberries, in Colorado, 42, 115, 146; in Kansas, 27, 122; Pathfinder and Trailblazers (cultivars), 207; in South Dakota, 163, 165; in Wyoming, 88, 89, 207, 208
Reclamation Act, 4, 54, 171–173
Redbud, in Kansas, 122
Report on the Lands of the Arid Region of the United States (Powell), 51–52, 53–58
Republican River: in Colorado, 107; Compact, 238; in Nebraska, 44
Rhubarb, 27; in Colorado, 42; in Wyoming, 208
Richards, William A., 92
Robertson, J. B., 113–114
Robertson, John Stevenston, 165
Rockafellow, B. F., 28
Rockmount Nursery, Boulder, 205

Rocky Ford, Colorado, 28, 147
Rocky Mountain News (newspaper), horticultural column, 231. *See also* Byers, William N.
Rocky Mountain Seed Company, 178, *178, 179,* 227, 230
Roosevelt, Eleanor, 195
Roosevelt, Franklin D., 185, 195
Roosevelt, Theodore, 4, 75, 80, 172, 174, 175, 185. *See also* Country Life Commission
Rutabagas, in Nebraska, 187; in Wyoming, 89, 208, 217
Rydberg, Per Axel, 70, 72–73, 78–79, 148

St. Vrain River, Colorado, 14; valley of, 28, 59
Salsify, in Nebraska, 187
Sandhills, Nebraska, 70, 72, 73–75, 79, 245
Sandoz, Jules, 13, 111, 163, 187
Sandsten, Emil Peter, 179
Savage, Ezra P., 75
Schaal, Lawrence A., 220
Schlyer, J., 121
Science, applications of, 64–68, 71, 176–177
Scott, Charles A., 75, 124
Scottsbluff, Nebraska, 187–188
Seeds, 14; collection of, 121, 205; Congressional distribution of, 146–147; laws on pure, 178–179; regional sales of, 146–147, 177, 178
Serviceberry, in Colorado, 145; in Wyoming, 88
Shelterbelts. *See* Windbreaks
Smith-Lever Act, 4, 103, 173, 174–175
Sod: "busting," 12, 89, 113; compacted by buffalo, 53; houses built of, *13,* 117
Soil conservation districts, 186–189
Sopris, Richard, 139–140
South Dakota: Agricultural College (South Dakota State University), 156, 165; agricultural experiment station, 151, 152, 157, 159, 166, 167; Horticultural Society, 22, 151, 165, 244
Species. *See* Plants
Speer, Robert W., 140–145, 175
Spinach, in Nebraska, 188; in Wyoming, 208, 214
Spruce: Colorado blue, in Nebraska, 78, 185; Colorado blue, in Wyoming, 219;

Koster's blue, in Nebraska, 185; white, in Nebraska, 185
Squashes, 14; in Colorado, 114; in Kansas, 120; in Nebraska, 187; in Wyoming, 208, 217. *See also* Pumpkin
Stanley, W. E., 121
Stegner, Wallace, 54
Stephens. E. F., 111
Stephens, John M., 197–198, 199, 201
Sterling, Colorado, 173, 176
Stolley, William, 22
Strawberries, 208; in Colorado, 42, 117, 146, 148; Fort Laramie (cultivar), 207; in Kansas, 27, 122; Ogallala (cultivar), 10, 207; in Kansas, 27, 122; in South Dakota, 163, 165; in Wyoming, 207
Street, W. D., 120
Sycamore, in Denver, 145

Talbot, John, 47
Taylor, William A., 202
Thistles, 71–72, 91–92, 148
Tillers, rotary, 180
Timber Culture Act, 19, 23, 27, 52, 54, 74, 116
Toliver, J. C., 73–74
Tomatoes: Bison (cultivar), 188; Danmark (cultivar), 209; Highlander (cultivar), 210; in Kansas, 120, 121, 209, 210; in Nebraska, 79, 187, 188; in Wyoming, 208, 209–210, *211,* 217
Torrington, Wyoming, 187, 234
Tree City USA, designation of, 235, 246
Trees and shrubs, viii, 13, 137, 146, 148; winter-kill of, 92, 122. *See also* Cheyenne Horticultural Field Station; Clarke-McNary Act; Colorado; Kansas; Nebraska; Wyoming; *individual varieties by name*
True, Alfred E., 115
Turnips: in Kansas, 120; in Nebraska, 187; prairie, 8, 9; in Wyoming, 46, 89, 208, 217

Union Colony of Colorado (Greeley), 3, 31, 33–39, 51. *See also* Greeley
Universities, land-grant, 16, 60, 64, 81, 86, 175. *See also individual universities by name*
U.S. Department of Agriculture (USDA), 16, 156, 157, 205; Bureau of Forestry, 74; Bureau of Plant Industry, 177, 181, 202, 218–219, 221; Division of Botany, 109,

118; Office of Dry-Land Agriculture, 117, 118, 202; Office of Experiment Stations, 111; Office of Farm Management, 123; Office of Horticultural Crops, 202; Section of Seed and Plant Introduction, 157, 209. *See also* U.S. Forest Service
U.S. Forest Service, 74, 180; Bessey nursery, 72, *76, 77, 78, 79*, 180, 184, 201, 219; Nebraska National Forest, 19, 72, 75, 236
Utopianism, 34–35, 40, 135

Valentine, Nebraska, 159
Variety. *See* Plants
Vasey, George, 109
Vegetables, heirloom, lists of, 120, 186–187, 214, 217. *See also* Gardens, vegetable; *individual varieties by name*
Vetting, Frederick C., 177–178
Vetting, Kenneth, 228
Victory Gardens, 175, *176*, 189, 210. *See also* Gardens, urban
Vines, native, 94–96

Wagner, Mrs. Adam S., 189
Wall, David K., 14
Wallace, Kansas, 26
Walnut, in Colorado, 117; in Kansas, 120, 123; in Nebraska, 70, 185
War: Civil War, 3, 11, 28, 47; World War I, 175; World War II, 180, 189, 210
Ward, Mrs. Jasper, D., 139
Warren, Francis E., 171. 172, 196–199
Washburn, Mrs. A. L., 43
Water: beneficial use, 58–59; conservation of, 51; farmstead reservoirs of, 116–117, 120; management of, 59, 238; doctrine of prior appropriation of, 57–59; doctrine of riparian rights to, 57–58; as property of state, 59; rights to, 38, 41, 58; scarcity of, 44, 55, 171, 223. *See also* Denver; Groundwater; Irrigation; Pathfinder Reservoir
Watermelons: in Nebraska, 79; in Wyoming, 89
Watson, Larry, 228
Weaver, Arthur J., 185
Webb, Walter Prescott, 2, 16, 24, 53, 109
Webber, Herbert J., 69, 70

Weeds, definition of, 91–92
Welch, E.S. v. State Board of Horticulture, 102
Wenger, Leon R., 190–191
Werner, H. O., 209
Western Kansas Agricultural Research Center (WKARC), 75, 121–124, *122*, 179, 218. *See also* Kansas State University
Wheatland, Wyoming, 99, 172, 184, 235; agricultural experiment station, 88–89, 90
Wheatridge, Colorado, 146
White River, South Dakota, 2
Whitetop, 208
Wilber, Charles Dana, 26, 52–53
Wilcox, Lucius, 244. *See also* Field and Farm
Willows, 24; in Kansas, 28; in Wyoming, 89
Wilmore Nursery, W. W., 146, 205
Wilson, Kansas, 26
Wilson, Edward O., 244
Wilson, James ("Tama Jim"), 60, 90, 107, 147, 156, 157
Wilson, James Wilbur, 157
Wilson, Robert, 201, 202
Wind, 24, 72–73, 79, 89, 113, 114, 196
Windbreaks, 219; in Colorado, 115, *118*, 181, 182; definition of, 24–25; in Kansas, 123; in Nebraska, 184; in Wyoming, *182, 184*. *See also* Trees; U.S. Forest Service; Wind
Windmills, *25*, 44–46, *45*, 78, 117
WKARC. *See* Western Kansas Agricultural Research Center
Wolff, Joseph, 28
Wyoming, 85; Horticulture Act, 96–97, 102–103; legislature, 46, 59, 90, 92; State Board of Horticulture, 99–100, 102–103, 174; tree planting in, 93
Wyoming Stock Growers Association, 46
Wyoming, University of, 87–88, 184, 220; agricultural experiment station, 88, 94; Cooperative Extension Service, 183, 187, 197; Rocky Mountain Herbarium, 94. *See also* Archer; Buffum; Cheyenne; Nelson, Aven; Wheatland

Xeriscaping, 140, 191, 228–230, *229*. *See also* Farming, dry-land

Yankton, South Dakota, 151
Yeager, Albert F., 188

www.ingramcontent.com/pod-product-compliance
Lightning Source LLC
Chambersburg PA
CBHW060553080526
44585CB00013B/547